EARTH OBSERVATIONS FROM SPACE

THE FIRST 50 YEARS OF SCIENTIFIC ACHIEVEMENTS

Committee on Scientific Accomplishments of Earth Observations from Space

Board on Atmospheric Sciences and Climate

Division on Earth and Life Studies

NATIONAL RESEARCH COUNCIL
OF THE NATIONAL ACADEMIES

THE NATIONAL ACADEMIES PRESS
Washington, D.C.
www.nap.edu

THE NATIONAL ACADEMIES PRESS 500 FIFTH STREET, N.W. WASHINGTON, DC 20001

NOTICE: The project that is the subject of this report was approved by the Governing Board of the National Research Council, whose members are drawn from the councils of the National Academy of Sciences, the National Academy of Engineering, and the Institute of Medicine. The members of the committee responsible for the report were chosen for their special competences and with regard for appropriate balance.

Support for this project was provided by the National Aeronautics and Space Administration under Contract No. NNG06GF62G. Any opinions, findings, conclusions, or recommendations expressed in this publication are those of the author(s) and do not necessarily reflect the views of the organizations or agencies that provided support for the project.

International Standard Book Number-13: 978-0-309-11095-2
International Standard Book Number-10: 0-309-11095-5

Additional copies of this report are available from the National Academies Press, 500 Fifth Street, N.W., Lockbox 285, Washington, DC 20055; (800) 624-6242 or (202) 334-3313 (in the Washington metropolitan area); Internet, http://www.nap.edu.

Cover design by Van Nguyen, National Academies Press.

Printed in the United States of America

THE NATIONAL ACADEMIES

Advisers to the Nation on Science, Engineering, and Medicine

The **National Academy of Sciences** is a private, nonprofit, self-perpetuating society of distinguished scholars engaged in scientific and engineering research, dedicated to the furtherance of science and technology and to their use for the general welfare. Upon the authority of the charter granted to it by the Congress in 1863, the Academy has a mandate that requires it to advise the federal government on scientific and technical matters. Dr. Ralph J. Cicerone is president of the National Academy of Sciences.

The **National Academy of Engineering** was established in 1964, under the charter of the National Academy of Sciences, as a parallel organization of outstanding engineers. It is autonomous in its administration and in the selection of its members, sharing with the National Academy of Sciences the responsibility for advising the federal government. The National Academy of Engineering also sponsors engineering programs aimed at meeting national needs, encourages education and research, and recognizes the superior achievements of engineers. Dr. Charles M. Vest is president of the National Academy of Engineering.

The **Institute of Medicine** was established in 1970 by the National Academy of Sciences to secure the services of eminent members of appropriate professions in the examination of policy matters pertaining to the health of the public. The Institute acts under the responsibility given to the National Academy of Sciences by its congressional charter to be an adviser to the federal government and, upon its own initiative, to identify issues of medical care, research, and education. Dr. Harvey V. Fineberg is president of the Institute of Medicine.

The **National Research Council** was organized by the National Academy of Sciences in 1916 to associate the broad community of science and technology with the Academy's purposes of furthering knowledge and advising the federal government. Functioning in accordance with general policies determined by the Academy, the Council has become the principal operating agency of both the National Academy of Sciences and the National Academy of Engineering in providing services to the government, the public, and the scientific and engineering communities. The Council is administered jointly by both Academies and the Institute of Medicine. Dr. Ralph J. Cicerone and Dr. Charles M. Vest are chair and vice chair, respectively, of the National Research Council.

www.national-academies.org

COMMITTEE ON SCIENTIFIC ACCOMPLISHMENTS OF EARTH OBSERVATIONS FROM SPACE

JEAN BERNARD MINSTER (*Chair*), Scripps Institution of Oceanography, La Jolla, California
JANET W. CAMPBELL (*Vice Chair*), University of New Hampshire, Durham
JEFF DOZIER, University of California, Santa Barbara
JAMES R. FLEMING, Colby College, Waterville, Maine
JOHN C. GILLE, National Center for Atmospheric Research, Boulder, Colorado
DENNIS L. HARTMANN, University of Washington, Seattle
KENNETH JEZEK, The Ohio State University, Columbus
STANLEY Q. KIDDER, Colorado State University, Fort Collins
NAVIN RAMANKUTTY, McGill University, Montreal, Quebec
ANNE M. THOMPSON, Pennsylvania State University, University Park
SUSAN L. USTIN, University of California, Davis
JAMES A. YODER, Woods Hole Oceanographic Institution, Massachusetts

NRC Staff

CLAUDIA MENGELT, Study Director
MARIA UHLE, Program Officer
LEAH PROBST, Research Associate
KATHERINE WELLER, Senior Program Assistant

BOARD ON ATMOSPHERIC SCIENCES AND CLIMATE

F. SHERWOOD ROWLAND (*Chair*), University of California, Irvine
M. JOAN ALEXANDER, NorthWest Research Associates, Boulder, Colorado
MICHAEL L. BENDER, Princeton University, Princeton, New Jersey
ROSINA M. BIERBAUM, University of Michigan, Ann Arbor
CAROL ANNE CLAYSON, Florida State University, Tallahassee
WALTER F. DABBERDT, Vaisala, Inc., Boulder, Colorado
KERRY A. EMANUEL, Massachusetts Institute of Technology, Cambridge
DENNIS L. HARTMANN, University of Washington, Seattle
PETER R. LEAVITT, Weather Information, Inc., Newton, Massachusetts
VERNON R. MORRIS, Howard University, Washington, D.C.
THOMAS H. VONDER HAAR, Colorado State University/CIRA, Fort Collins

Ex Officio Member

ANTONIO J. BUSALACCHI, JR., University of Maryland, College Park

NRC Staff

CHRIS ELFRING, Director
IAN KRAUCUNAS, Program Officer
CURTIS H. MARSHALL, Program Officer
CLAUDIA MENGELT, Program Officer
ELIZABETH A. GALINIS, Research Associate
LEAH PROBST, Research Associate
ROB GREENWAY, Senior Program Assistant
KATHERINE WELLER, Senior Program Assistant
DIANE GUSTAFSON, Administrative Coordinator
SHUBHA BANSKOTA, Financial Associate

Preface

Over the past five decades, space-faring nations have developed impressive capabilities for observing Earth from satellite platforms. This has resulted in rapid advances in fundamental science and improved our ability to understand and predict the dynamics of Earth systems, to the great benefit of society. Global geophysical observations covering a wide range of disciplines have provided unprecedented insight into the physics of Earth systems. Exquisitely accurate space geodetic measurements have yielded a global reference system that is three orders of magnitude more accurate than that of a half-century ago. Today, our ability to forecast weather, climate, and natural hazards depends critically on satellite-based observations of the planet.

The Earth science community is currently engaged in major efforts to plan directions for future observations and research that depend on space-based platforms. One essential study—the first of its kind in the Earth sciences—is the recent "decadal survey" led by the National Academies: *Earth Science and Applications from Space: National Imperatives for the Next Decade and Beyond* (NRC 2007a). This is a forward-looking report that provides guidance to the U.S. government—particularly the National Aeronautics and Space Administration (NASA), the U.S. Geological Survey, and the National Oceanic and Atmospheric Administration—about future priorities. It recommends a renewal of the national commitment to support ongoing observations from space in order to face scientific and societal challenges over the next decades and to understand and manage natural resources. To complement this decadal survey, NASA asked the National Academies to illustrate the value of a half-century of Earth observations from space. That is the topic of this report.

The committee addressed this task by meeting with expert scientists from various disciplines who offered their perspectives on crucial discoveries and scientific achievements enabled by satellite observations. Suggestions were also solicited from the Earth science community at large through various distribution lists, and a town hall meeting was organized through the American Geophysical Union to elicit comments and opinions. From this process emerged a long list of scientific accomplishments, each singularly convincing, unique, and sometimes amazing that could not have been achieved without orbital observations. From this list the committee chose some of the most compelling and illustrative examples to showcase the value of satellite observations and argue the central importance of sustaining the effort to develop and deploy these observational tools. The examples presented in this report capture the committee's subjective view of the most important accomplishments, yet we believe that another committee's short list would overlap considerably with the present selection. These accomplishments demonstrate clearly that the advent of satellite observations has revolutionized the Earth sciences.

Many individuals contributed essential information and helped in writing the examples of accomplishments (Chapters 3-11) and added to the committee members' own expertise, and they are recognized in the Acknowledgments. I am also grateful to the committee members; they volunteered countless hours to this study. I would like particularly to acknowledge the dedication of Janet Campbell, the committee's vice chair, who "stepped into the breach" more than once to keep our momentum.

Finally, I am most grateful to the National Research Council (NRC) staff—Study Director Claudia Mengelt; Research Associate Leah Probst, who managed the circulation of drafts and supporting documents via the Web and staffed all of our meetings; Maria Uhle, who pitched in during the busiest periods of the study; and Senior Program Assistant Katie Weller, who handled the logistics of all meetings, cheerfully managing the travel requirements of all participants. As is usual for NRC studies, the staff was a critical element in completing the study on time and on budget.

Bernard Minster, *Chair*
Committee on Scientific Accomplishments of Earth Observations from Space

Acknowledgments

Over the course of this study, the committee met five times to gather information and conduct deliberations. In the process, many members of the Earth science community were invited to provide input and contribute descriptions of scientific accomplishments. In addition, the community responded to a broad solicitation for input, which also helped shape the committee's thinking. In particular, the committee wants to acknowledge the following individuals for providing invaluable information to this study:

Tad Anderson, University of Washington, Seattle
Rick Anthes, University Corporation for Atmospheric Research, Boulder, Colorado
Robert Bindschadler, National Aeronautics and Space Administration (NASA)
Kenneth Casey, National Oceanic and Atmospheric Administration (NOAA)
Yi Chao, NASA
Dudley Chelton, Oregon State University, Corvallis
James Coakley, Oregon State University, Corvallis
Peter Cornillon, University of Rhode Island, Narragansett
Diane Evans, NASA
Jonathan A. Foley, University of Wisconsin, Madison
Mark Friedl, Boston University, Massachusetts
Randy Friedl, NASA
Ralph Kahn, NASA
Jack Kaye, NASA
Dennis Lettenmaier, University of Washington, Seattle
Ulrike Lohmann, Institute for Atmospheric and Climate Science, Eidgenössische Technische Hochschule, Zurich, Switzerland
Stephen Lord, NOAA
Tom Loveland, U.S. Geological Survey
Michael Mishchenko, NASA
Lorraine Remer, NASA
Sassan Saatchi, NASA
Annemarie Schneider, University of California, Santa Barbara
Bill Smith, Hampton University, Virginia
Dave Smith, NASA
Omar Torres, NASA
Lucia Tsaoussi, NASA
Bruce Wielicki, NASA
Carl Wunsch, Massachusetts Institute of Technology, Cambridge
Charles Yentsch, Bigelow Laboratory for Ocean Sciences, West Boothbay Harbor, Maine
Howard Zebker, Stanford University, California

This report has been reviewed in draft form by individuals chosen for their diverse perspectives and technical expertise, in accordance with procedures approved by the National Research Council's Report Review Committee. The purpose of this independent review is to provide candid and critical comments that will assist the institution in making its published report as sound as possible and to ensure that the report meets institutional standards for objectivity, evidence, and responsiveness to the study charge. The review comments and draft manuscript remain confidential to protect the integrity of the deliberative process. We wish to thank the following individuals for their review of this report:

Gregory Asner, Carnegie Institution, Stanford, California
Sheldon Drobot, University of Colorado, Boulder
Dara Entekabi, Massachusetts Institute of Technology, Cambridge
Inez Fung, University of California, Berkeley
Bradford Hager, Massachusetts Institute of Technology, Cambridge
Charles Kolb, Aerodyne Research, Inc., Billerica, Massachusetts
Kuo-Nan Liou, University of California, Los Angeles

Anne Nolin, Oregon State University, Corvallis
David Siegel, University of California, Santa Barbara
Lynn Talley, Scripps Institution of Oceanography, La Jolla, California
Byron Tapley, University of Texas, Austin
Thomas Vonder Haar, Colorado State University, Fort Collins

Although the reviewers listed above provided constructive comments and suggestions, they were not asked to endorse the report's conclusions or recommendations, nor did they see the final draft of the report before its release. The review of this report was overseen by Elbert W. Friday, Jr., University of Oklahoma, Norman. Appointed by the National Research Council, he was responsible for making certain that an independent examination of this report was carried out in accordance with institutional procedures and that all review comments were carefully considered. Responsibility for the final content of this report rests entirely with the authoring committee and the institution.

Contents

Summary

Just as the invention of the mirror allowed humans to see their own image with clarity for the first time, Earth observations from space have allowed humans to see themselves for the first time living on and altering a dynamic planet.

Observing Earth from space over the past 50 years has fundamentally transformed the way people view our home planet. The image of the "blue marble" (Figure S.1) is taken for granted now, but it was revolutionary when taken in 1972 by the crew on Apollo 17. Since then the capability to look at Earth from space has grown increasingly sophisticated and has evolved from simple photographs to quantitative measurements of Earth properties such as temperature, concentrations of atmospheric trace gases, and the exact elevation of land and ocean. Imaging Earth from space has resulted in major scientific accomplishments; these observations have led to new discoveries, transformed the Earth sciences, opened new avenues of research, and provided important societal benefits by improving the predictability of Earth system processes.

This report highlights the scientific achievements made possible by the first five decades of Earth satellite observations by space-faring nations. It follows on a recent report from the National Research Council (NRC) entitled *Earth Science and Applications from Space: National Imperatives for the Next Decade and Beyond,*[1] also referred to as the "decadal survey." Recognizing the increasing need for space observations, the decadal survey identifies future directions and priorities for Earth observations from space. This companion report was requested by the National Aeronautics and Space Administration (NASA) to highlight, through selected examples, important past contributions of Earth observations from space to our current understanding of the planet.

[1]National Research Council (NRC). 2007. Earth Science and Applications from Space: National Imperatives for the Next Decade and Beyond. The National Academies Press, Washington, D.C.

A UNIQUE VANTAGE POINT

The 1957-1958 International Geophysical Year (IGY), with 67 participating nations, was an unprecedented effort referred to by noted geophysicist Sydney Chapman (1888-1972) as "the common study of our planet by all nations for the benefit of all." Teams of observers were deployed around the globe—some to the ends of the Earth in polar regions, on high mountaintops, and at sea—to study Earth processes. The effort in Antarctica alone involved hundreds of people in logistically complex and expensive expeditions. Even in 1957 it was recognized that satellite data would provide observations of the Earth system that no amount of ground-based observations could achieve. During the IGY the Soviet Union launched the world's first satellite, Sputnik, in October 1957 and transformed the Earth science endeavor. Shortly thereafter, the United States launched its first satellite, Explorer 1, in January 1958. Over the course of the next five decades, an array of satellites have been launched that have fundamentally altered our understanding of the planet. Today from the comfort of their desks, Earth scientists can acquire global satellite data with orders of magnitude greater spatial and temporal coverage than obtained during the intensive field expeditions of the IGY.

The global view obtained routinely by observations from space is unmatched in its ability to resolve the dynamics and the variability of Earth processes. Ship-based observations, for example, cannot provide the spatial and temporal information to detect the dynamic nature of the ocean. Similarly, aircraft and weather balloon measurements alone cannot resolve the details required to understand the complex dynamics of ozone depletion. Space observations provide detailed quantitative information on many atmospheric, oceanic, hydrologic, cryospheric, and biospheric processes. Because satellite information is gathered at regular intervals, it provides, like a movie, a view of changes over time. For the first time, satellites make it possible to track a tropical

FIGURE S.1 The blue marble as seen by the crew of Apollo 17. Image (AS17-148-22727) courtesy of the Image Science & Analysis Laboratory, NASA Johnson Space Center. SOURCE: *http://eol.jsc.nasa.gov*.

cyclone from its gestation over the ocean to landfall and to observe the ever-fluctuating intensity of the storm.

Over the past two decades, this dynamic global view has radically transformed our understanding of ice sheets. Before satellites, Antarctica's and Greenland's ice sheet mass balance was assumed to be controlled by the difference between ice melting and accumulation rates, and the rate of ice discharge into the ocean was assumed to be constant. Satellite radar images from RADARSAT revealed that (1) the velocity of ice sheet flow is highly variable, (2) there exist complex networks of ice streams, and (3) the velocity of ice stream flow toward the sea has increased measurably in response to climate change. The collapse of the Larsen B Ice Shelf in Antarctica in 2002—captured only because of frequent coverage by satellite imagery—dramatically illustrated the dynamics of ice sheets on astonishingly short timescales (Figure S.2). These revelations carry weighty implications: the rapid transfer of ice from the continental ice sheets to the sea could result in a significant rise of sea level.

One of the most effective ways to illustrate the impact that observations from space have had on weather forecasting is to watch the weather in motion in a sequence of satellite images (e.g., *http://www.goes.noaa.gov*). These images are captured by geostationary satellites, first positioned over the equator in the mid-1960s. They collect frequent photographs at various wavelengths, from which the moving pictures of weather can be assembled. Geostationary satellites rapidly became a major source of data for weather services worldwide, which is now essential to air traffic management, disaster preparedness, agriculture, and many other everyday applications.

FUNDAMENTAL CONTRIBUTIONS TO SCIENCE

This report describes many examples of scientific accomplishments from satellite observations that have transformed the Earth sciences, some of which are highlighted in this summary (Table S.1). Few are as transformative as the advances in space geodesy over the past five decades, particularly with the ubiquitous introduction of Global Positioning System (GPS) devices, which have brought geodetic positioning to everyday life. At the time of the IGY, the geolocation of most points at the surface of Earth entailed errors that reached hundreds of meters in remote areas. Today, scientists rely on an International Earth Reference Frame from which geographical positions can be described relative to the geocenter, in three-dimensional Cartesian coordinates to centimeter accuracy or better—a two to three orders-of-magnitude improvement compared to 50 years ago. This is even more remarkable considering it is accomplished on an active planet whose surface is constantly in motion. The change in position of the rotation axis (the poles) is determined daily to centimeter accuracy, and the changes in length of day are determined to millisecond accuracy within a few hours. Inexpensive GPS receivers are now taken for granted by consumers who are rapidly becoming accustomed to GPS navigation on the road, on the water, or in the air without realizing the enormous body of science behind this technological achievement: accurate position information of the satellites, very stable clocks, and well-calibrated atmospheric corrections.

Satellite-derived global maps of air pollution caused a major change in concepts of pollution control by demonstrating its transport between nations and continents that. The first tropospheric ozone maps from space in the 1980s drew attention to human impacts on the atmosphere, especially in the tropics where agricultural fires and land-use changes alter ozone in the lower atmosphere. Newer satellites show plumes of ozone, aerosols, and gases such as carbon monoxide spanning oceans and linking continents. Therefore, pollution is now viewed as a global, not a local, phenomenon. Quantitative information from satellites provides data for modeling efforts to predict coupled atmospheric chemical and climate changes with greater confidence. Orbital sensors precisely locate sources of ozone-destroying bromine monoxide (from bromide in sea ice and sea salt particles) and nitrogen and sulfur oxides (from urban regions, power plants, and smelting operations). In combination with numerical models, the global sources of these gases can be mapped and tracked.

Climate science has also advanced spectacularly through satellite observations. The radiometer flown on Explorer 7 from 1959 to 1961 made it possible for the first time to directly measure the energy entering and leaving Earth. This and follow-on missions enabled scientists to measure Earth's energy balance with much greater confidence

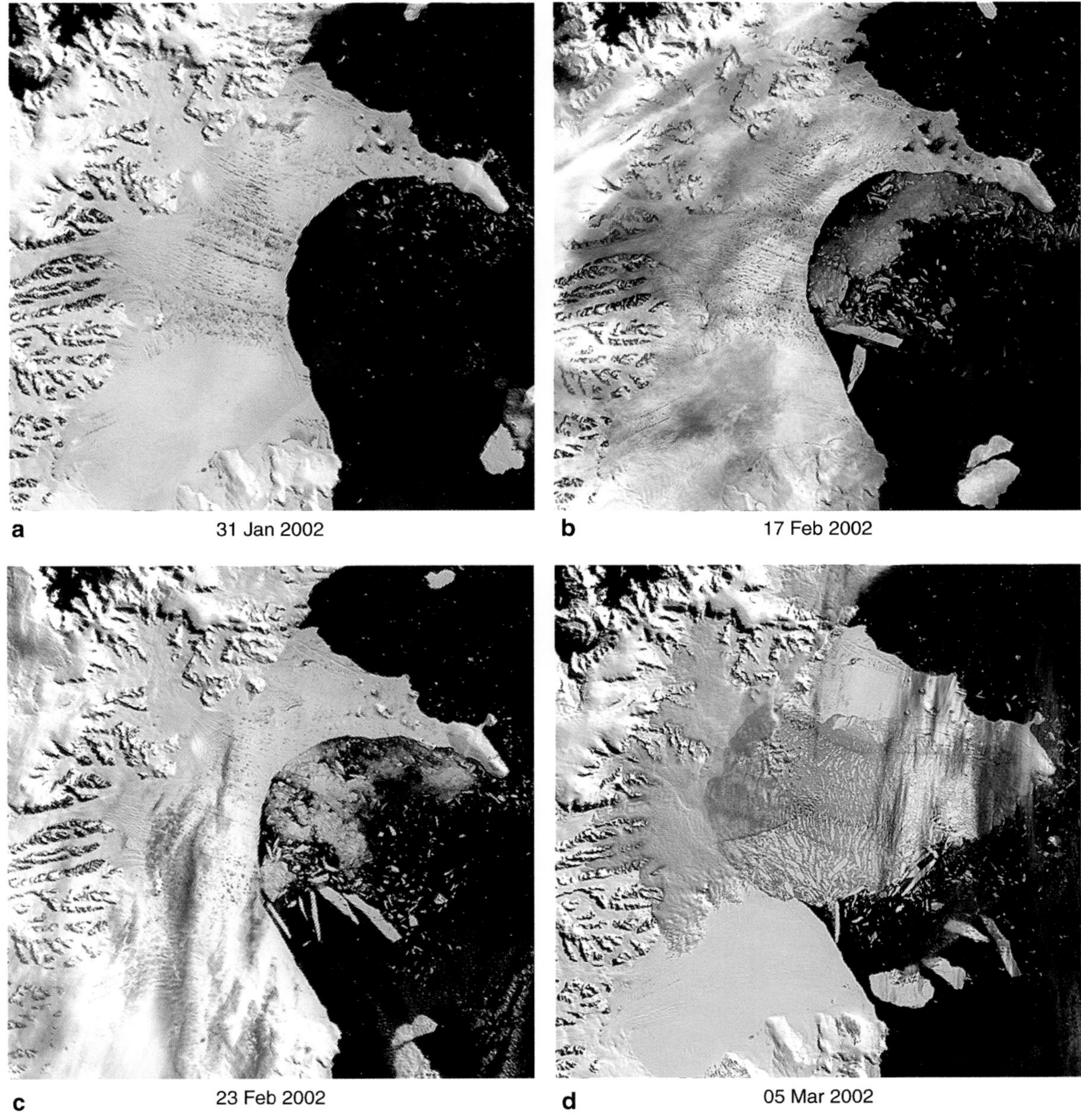

FIGURE S.2 Collapse of the Larsen B Ice Shelf in Western Antarctica, January-March 2002. Two thousand square kilometers of the ice shelf disintegrated in just 2 days. SOURCE: National Snow and Ice Data Center.

TABLE S.1 Examples of the Scientific Accomplishments of Earth Observations and Landmark Satellites That Have Contributed to Each

Accomplishment	Satellite
Monitoring global stratospheric ozone depletion, including Antarctica and Arctic regions	TIROS series, Nimbus 4 and 7, ERS 1, Envisat
Detecting tropospheric ozone	Nimbus 7, ERS 2, Envisat, Aqua, Aura, MetOp
Measuring the Earth's radiation budget	Explorer 7, TIROS, and Nimbus
Generating synoptic weather imagery	TIROS series, ATS, SMS
Assimilating data for sophisticated numerical weather prediction	Numerous weather satellites, including the TIROS series and NOAA's GOES and POES
Discovering the dynamics of ice sheet flows in Antarctica and Greenland	RADARSAT, InSAR, Landsat, Aura, and Terra
Detecting mesoscale variability of ocean surface topography and its importance in ocean mixing	TOPEX/Poseidon
Observing the role of the ocean in climate variability	TIORS-N and NOAA series
Monitoring agricultural lands (a contribution to the Famine Early Warning System)	Landsat
Determining the Earth reference frame with unprecedented accuracy	LAGEOS, GPS

compared to earlier indirect estimates resulting in improved climate models. Over the years, as radiometers improved, these measurements achieved the precision, spatial resolution, and global coverage necessary to observe directly the perturbations in Earth's global energy budget associated with short-term events such as major volcanic eruptions or the El Niño-Southern Oscillation (ENSO). In addition, radiometers in orbit nearly continuously since the 1960s directly measure the equator-to-pole heat transport by the climate system, the greenhouse effect of atmosphere trace gases, and the effect of clouds on the energy budget of Earth. These observations advance our understanding of the climate system and improve climate models.

Another important contribution to climate science was made by the long-term record of sea surface temperature (SST) from the Advanced Very High Resolution Radiometer (AVHRR) flown on the Television Infrared Observation Satellite series (TIROS-N) and the National Oceanic and Atmospheric Administration (NOAA) satellite series. As the longest oceanographic data record from remote sensing, it had broad impact. The SST record exposed the role of the ocean in regional and global climate variability and revealed important details about ocean currents. Trend analysis of the SST record provided evidence for global warming as 80 percent of the excess heat is entering the ocean and also helped improve understanding of the important climate-atmosphere feedbacks in the tropics that are also responsible for ENSO events. Understanding the increase in SST and anthropogenic heat input to the surface ocean also has important ramifications for quantifying and predicting sea-level rise in response to global warming.

Very accurate measurements of sea surface heights by the Topographic Experiment (TOPEX)/Poseidon altimeter have revolutionized our understanding of ocean dynamics. These observations allow scientists to characterize the scales and energy of mesoscale[2] features at a global scale and thus have revolutionized our understanding of basin-scale interannual variability, such as El Niño events. Altimetry observations also improved our understanding of mean ocean circulation. The newly discovered prevalence of ocean eddies revolutionized the way oceanographers think about the mesoscale energy sources for deep-ocean mixing. The new paradigm is that of a very dynamic, turbulent system, with the energy primarily provided by winds and tides that are variable on many timescales.

[2]In the size range of 10-100 km.

SOCIETAL APPLICATIONS OF SATELLITE DATA

The most broadly used products from satellites are weather observations that enable forecasts. Since satellite images have become readily available, no tropical cyclone (hurricane or typhoon) has gone undetected, which provides affected coastal areas with advance warning and crucial time to prepare. This exemplifies not only how satellite observations have transformed the Earth sciences but also how the improved predictability of Earth processes can provide direct societal benefits. Weather forecasts more than a few hours into the future are made with the aid of numerical weather prediction models. By assimilating satellite observations, which yield dramatically improved and continually updated knowledge of the state of the atmosphere, meteorologists can devise models that project the weather into the future with much improved accuracy compared to presatellite forecasts. Consequently, 7-day forecasts have improved significantly in accuracy over the past decades, particularly for the relatively data-sparse southern hemisphere.[3] Needless to say, these improvements in forecast skills are saving countless human lives and have an enormous economic value (saving the energy sector alone hundreds of millions of dollars).

The ability to detect land-cover changes at all spatial

[3]Anomaly correlation of 500 hPa height forecasts for medium-range forecasts improved from 30 to 70 percent in the southern hemisphere (~45 to 70 percent in the northern hemisphere) between 1981 and 2006 (Simmons and Hollingsworth 2002).

scales from space has also produced extraordinary societal benefits. The phenomenal advantage brought by satellite information in monitoring and, more importantly, enabling forecasts of the productivity of large-area crops was demonstrated in the early 1970s. Since then federal agencies such as the U.S. Department of Agriculture have routinely used multispectral satellite imagery—offered by the Landsat series and other missions—in crop commodity forecasting. A particularly noteworthy application is the Famine Early Warning System Network, which was initially set up in Sub-Saharan Africa and now operates in other arid environments of the developing world. This system uses satellite images in conjunction with ground-based information to predict and mitigate famines.

Another example of an important societal benefit gained from satellite observations is the continuous observation of stratospheric ozone. Satellite observations from the Nimbus series (1980s) provided the first global maps of ozone depletion by man-made chlorine- and bromine-containing compounds released to the atmosphere. These observations combined with field studies, were critical to the development and adaptation of the Montreal Protocol, opened to signatures in 1987, and subsequent amendments designed to phase out ozone-destroying halogenated compounds. Since then satellite observations from newer platforms (Aura series) have continued to verify model predictions of the amounts and distributions of the causative agents. They also track variations in the size and depth of the annual Antarctic ozone hole (Figure S.3) and the more subtle but dangerous losses of ozone over heavily populated midlatitudes. Recent satellite observations show a decrease in chlorine-containing gases and the apparent beginning of an ozone recovery at midlatitudes, yielding increased confidence that the Montreal Protocol is indeed achieving its goal.

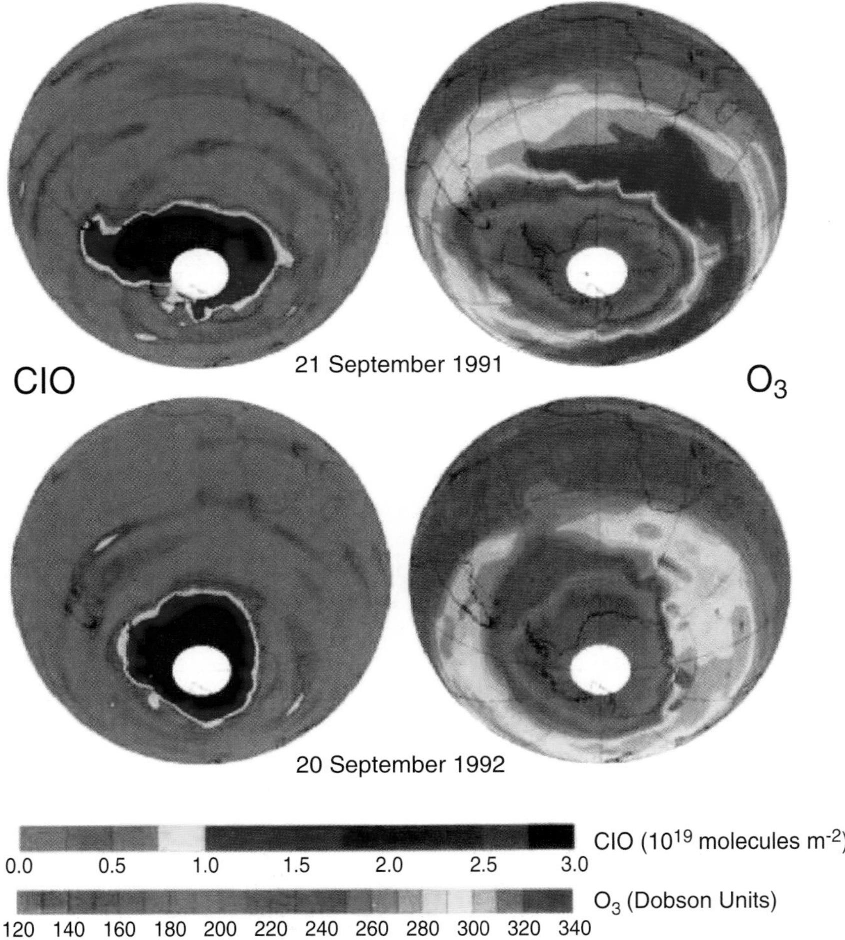

FIGURE S.3 Chlorine monoxide (ClO; left panel) and stratospheric ozone (O_3; right panel) columns over the southern hemisphere measured by the Microwave Limb Sounder (MLS) on the Upper Atmosphere Research Satellite (UARS) for days during the austral springs of 1991 and 1992. These images show that high ClO concentrations coincide in space and time with low O_3 concentrations confirming ground-based measurements and the proposed mechanisms for ozone depletion. The white circle over the pole indicates the area where no data is available. SOURCE: Waters et al. (1993). Reprinted with permission from Macmillian Publishers Ltd., copyright 1993.

INFRASTRUCTURE REQUIREMENTS TO ADVANCE SCIENCE

Earth observations from space demonstrate the successful synergy between science and technology. As scientists have gained experience in studying Earth through satellite observations, they have defined new technological needs, helped drive technological development to provide more quantitative and accurate measurements, and have advanced more sophisticated methods to interpret satellite data. To capitalize fully on the investment made in Earth-orbiting observing platforms and make the best use of these observations, satellite data require careful calibration and sophisticated analysis and assimilation tools. Optimal data processing can be undertaken only if a suitably trained workforce is in place to develop these tools and interpret the observations. In this respect, full and open access to satellite data is crucial because training and maintaining the required workforce is possible only if the data are continuously accessible to the broad scientific community.

The concept of open data access was adopted by the IGY when establishing the World Data Center System 50 years ago, and it is even more meaningful today than at the time of the Cold War. This does not preclude commercialization of some aspects of useful data product development, but the portion of carefully calibrated low-level data that is properly a public good should be made available to all stakeholders at no more than the cost of reproduction. The Landsat story is a case in point: wholesale commercialization of the data led to a precipitous drop in their use for both scientific and commercial applications, which recovered upon return to the earlier open data access policy. Only when academic, government, and commercial scientists are given liberal access to the data, and when a sufficient number of scientists are trained in the effective use of these data, will the analysis tools mature to the benefit of all parties. Our 50-year experience with passive (e.g., optical) and active (e.g., radar or lidar [light detection and ranging]) surface imagery, weather satellites, and planetary field measurements shows that the maturation process of these tools requires decades.

CONCLUSIONS

The first 50 years of Earth observations from space imparted the fundamental lessons that everything—land, ocean, and atmosphere—is intricately intertwined and that the Earth is a complex and dynamic system. In addition, "each [satellite] mission taught scientists not only something new about the Earth system, but also something new about how to create, operate, and improve the technology for observing the Earth from space."[4]

[4]NASA Earth Observatory, *http://earthobservatory.nasa.gov/Study/Nimbus 1.*

Based on its review of important scientific accomplishments, the committee concludes the following (for a detailed description, see Chapter 12):

1. The daily synoptic global view of Earth, uniquely available from satellite observations, has revolutionized Earth studies and ushered in a new era of multidisciplinary Earth sciences, with an emphasis on dynamics at all accessible spatial and temporal scales, even in remote areas. This new capability plays a critically important role in helping society manage planetary-scale resources and environmental challenges.

2. To assess global change quantitatively, synoptic data sets with long time series are required. The value of the data increases significantly with seamless and intercalibrated time series, which highlight the benefits of follow-on missions. Further, as these time series lengthen, historical data sets often increase in scientific and societal value.

3. The scientific advances resulting from Earth observations from space illustrate the successful synergy between science and technology. The scientific and commercial value of satellite observations from space and their potential to benefit society often increase dramatically as instruments become more accurate.

4. Satellite observations often reveal known phenomena and processes to be more complex than previously understood. This brings to the fore the indisputable benefits of multiple synergistic observations, including orbital, suborbital, and in situ measurements, linked with the best models available.

5. The full benefits of satellite observations of Earth are realized only when the essential infrastructure, such as models, computing facilities, ground networks, and trained personnel, is in place.

6. Providing full and open access to global data to an international audience more fully capitalizes on the investment in satellite technology and creates a more interdisciplinary and integrated Earth science community. International data sharing and collaborations on satellite missions lessen the burden on individual nations to maintain Earth observational capacities.

7. Over the past 50 years, space observations of the Earth have accelerated the cross-disciplinary integration of analysis, interpretation, and, ultimately, our understanding of the dynamic processes that govern the planet. Given this momentum, the next decades will bring more remarkable discoveries and the capability to predict Earth processes, critical to protect human lives and property. However, the nation's commitment to Earth satellite missions must be renewed to realize the potential of this fertile area of science.

Because the critical infrastructure to make the best use of satellite data takes decades to build and is now in place, the scientific community is poised to make great progress toward understanding and predicting the complexity of the Earth system. However, building a predictive capability relies strongly on the availability of intercalibrated long-term data records, which can only be maintained if subsequent generations of satellite sensors overlap with their predecessors. As the decadal survey points out, the capability to observe Earth from space is jeopardized by delays in and lack of funding for many critical satellite missions.

The decadal survey and this committee both recommend that the nation's commitment to continue Earth observations from space be renewed. Resources will be required to maintain the current momentum and not risk losing the workforce and infrastructure built over the past decades. Given the many scientific challenges ahead, we have seen only the beginning of an era of Earth observations from space. A report in 50 years will present many more significant achievements and discoveries and highlight how satellites played a vital role in observing the dynamics of the Earth system and in guiding our nation and others in meeting the challenges posed by global changes.

1

Introduction

Over the past 50 years, the United States has developed impressive capabilities for observing Earth from space-based platforms. Global observations of a wide range of geophysical and biological parameters have provided unprecedented insight into how the Earth system functions and have led to many fundamental scientific advances. The capacity to forecast and project weather, climate, and environmental hazards has benefited extensively from satellite-based Earth observations. Satellite observations have literally transformed the way we view the planet. Earth observations from space have provided the information needed to verify and complete our understanding of how ozone chemistry in the stratosphere controls the infamous polar ozone hole; how ocean currents, temperatures, and atmospheric processes are coupled to the El Niño-Southern Oscillation; how snow cover affects water cycle dynamics; and how numerous global and regional factors influence sea level change.

The Earth science community is devoting significant efforts to planning future observations and research to be conducted with space-based platforms. One important element of this effort is the recent "decadal survey" led by the National Academies: *Earth Science and Applications from Space: Urgent Needs and Opportunities to Serve the Nation* (NRC 2005) and *Earth Science and Applications from Space: National Imperatives for the Next Decade and Beyond* (NRC 2007a). These are two forward-looking reports funded jointly by the National Aeronautics and Space Administration (NASA), U.S. Geological Survey, and National Oceanic and Atmospheric Administration (NOAA). The reports provide guidance to the three agencies regarding future priorities and briefly summarize the value of satellite observations to the well-being of society. However, the continued availability of environmental satellite data in the future is jeopardized because of budgetary constraints and programmatic difficulties (NRC 2005, 2007a). Consequently, the reports recommend a renewal of the commitment to support observations from space in view of the scientific and societal challenges of understanding and managing natural resources over the coming decades. In addition to the decadal survey, NASA asked the National Academies to look back on the history of space-based observations of Earth to illustrate the contributions to the scientific enterprise to date (Box 1.1).

THE STUDY'S APPROACH

The committee relied on the relevant literature, its own collective experience, and input from the Earth science community to compile an extensive list of accomplishments resulting from Earth observations from space. Using its expert judgment, the committee chose a subsample of the major accomplishments that are compelling and illustrative to convey the extent to which satellite observations have revolutionized the way people view, understand, and study Earth. The committee did not attempt to provide a comprehensive inventory of accomplishments resulting from satellite information and recognizes the inherent bias of composing any selective list. Nevertheless, the committee believes that any other group of Earth scientists would have provided a compilation with considerable overlap with the one presented in this report, in part because committee members were selected with careful consideration to balancing expertise.

In addition to its own expertise, the committee invited other experts to contribute in a variety of ways: the committee invited presentations during a series of meetings, held a town hall session at the 2006 Annual Meeting of the American Geophysical Union, posted a call for input on the committee's website, and approached the following Earth science communities with a solicitation for contributions on accomplishments from Earth satellite observations:

- members of the National Academy of Sciences and National Academy of Engineering,

BOX 1.1
Statement of Task

This study documents specific scientific accomplishments resulting from the nation's research and development of space-based Earth observational capabilities. The study committee sought broad community input to identify examples of important accomplishments, in part by drawing on the expertise of the various entities within the National Academies and also involving those scientific communities that develop and use remote sensing observations of the Earth.

The study's main objective is to document, using examples and explanation, how satellite observations uniquely contributed to scientific understanding of the atmosphere, ocean, land, biosphere, and cryosphere. As secondary objectives, the study also addresses how satellite observations have contributed to the ability to predict variations in the Earth system (e.g., weather, climate variability, water availability, earthquakes, volcanoes, and tsunamis) and comments on opportunities to improve future Earth science research enabled by the vantage point of space.

To the extent possible, the committee organizes its comments to correspond to NASA's seven Earth science foci: (1) atmospheric composition; (2) carbon cycle and ecosystems; (3) climate variability and change; (4) earth surface and interior structure; (5) weather; (6) water and energy cycles; and (7) Sun-Earth connection.

- members of relevant boards within the National Academies,
- recipients of the quarterly newsletter of the National Academies' Board on Atmospheric Sciences and Climate, and
- members of relevant scientific e-mail distribution lists.

This report begins with a brief early history of the evolution of Earth observations from space (Chapter 2). In subsequent chapters the committee presents examples of major scientific accomplishments that have transformed and contributed to the Earth sciences. The committee considered as major accomplishments only scientific advances that resulted in a new discovery, transformative science, proving or disproving an important theory, opening new major research venues, or providing significant societal benefits. These accomplishments are described in Chapters 3 to 11. In the final chapter (Chapter 12), the committee summarizes conclusions drawn from these major accomplishments and highlights opportunities to improve future Earth science research enabled from the vantage point of space.

2

Earth Observations from Space: The Early History

"Space technology affords new opportunities for scientific observation and experiment, which will add to our knowledge and understanding of the earth."
—President's Science Advisory Committee (1958)

The space age officially began on October 4, 1957, with the dramatic and historic launch of Sputnik 1 by the Soviet Union, but there are much deeper roots. Robert H. Goddard developed liquid-fueled rockets and used them for weather photography in the 1920s; remote sensing radio technology was developed in the 1930s and 1940s; a V-2 rocket flight in 1947 photographed clouds from an altitude of 100 miles; and by 1954 instrumented sounding rockets had serendipitously photographed an unknown tropical storm (Figure 2.1).

FIGURE 2.1 Image of a previously undetected tropical storm in the Gulf of Mexico photographed by an Aerobee sounding rocket in 1954. SOURCE: Hubert and Berg (1955). Reprinted with permission from the American Meteorological Society, copyright 1955.

In 1955 the United States announced it would launch a scientific Earth satellite during the International Geophysical Year (IGY) of 1957-1958. Sputnik, however, diverted the world's attention from scientific concerns and focused American perceptions on a "missile gap" and possible national security threats from space. The November launch of Sputnik 2 further fueled these fears. In response, and with the Navy's Vanguard Program languishing, the U.S. Department of Defense used a modified Redstone military missile, the Juno 1, to launch the first U.S. satellite, Explorer 1, on January 31, 1958.

That year Congress enacted the National Defense Education Act, which provided dramatically increased support for both basic research and science education at all levels and benefited the nation in subsequent generations in ways that could not have been foreseen at the time. The National Aeronautics and Space Act created a new agency—the National Aeronautics and Space Administration (NASA)—to consolidate and lead the U.S. space effort. Its mission included expanding knowledge of phenomena both within the atmosphere and in outer space and developing and operating vehicles capable of carrying instruments for peaceful and scientific purposes in cooperation with other nations. Also in 1958, the President's Science Advisory Committee pointed out that a satellite in orbit could be used for three scientific purposes: "(1) it can sample the strange new environment though which it moves; (2) it can look down and see the earth as it has never been seen before; and (3) it can look out into the universe and record information that can never reach the earth's surface because of the intervening atmosphere" (PSAC 1958).

Even in 1958, scientists knew that Earth-orbiting satellites would encounter and be able to measure the charged plasma of the solar wind, the three-dimensional structure of Earth's gravitational and magnetic fields, and other energy fluxes and fast-moving particles in near space. Satellites could detect incoming cosmic rays and be used to test Einstein's general theory of relativity; they could measure solar energy at the top of the atmosphere to determine how much is reflected and radiated back to space by clouds, oceans, continents, and ice sheets; and they could look down at Earth to expand surveillance of weather systems from about 10 percent to practically 100 percent of Earth's surface. The scientific accomplishments of five prominent early satellites (Explorer 1, Explorer 7, TIROS 1 [Television Infrared Observing Satellite], ATS 1 [Application Technology Satellite], and the Nimbus series), all launched in the first two decades of the space age, fulfilled many of these expectations.

EARLY SATELLITES AND PIONEERS

"At the dawn of the Space Age, the nature of space exploration was already apparent: It always leads to unexpected discoveries about our universe and the processes that shape our environment" (Friedman 2006). On May 1, 1958, University of Iowa scientist James Van Allen (Box 2.1, Figure 2.2) announced that Geiger-Müller counters aboard the Jet Propulsion Laboratory (JPL) Explorer 1 (Figure 2.3) and Explorer 3 satellites had been swamped by high radiation levels at certain points in their orbits, indicating that powerful radiation belts, later known as the Van Allen belts, surround Earth (Van Allen et al. 1958, Van Allen and Frank 1959). Vanguard 1, the fourth artificial satellite launched, provided important geodetic information about the shape of the Earth, specifically its north-south asymmetry (O'Keefe et al. 1960).

NASA launched the world's first weather satellite, TIROS 1, on April 1, 1960 (Figure 2.4). TIROS 1 flew in a nearly circular, prograde orbit of 48 degrees inclination. It took television (Figure 2.5) and (on later flights) infrared photos of weather patterns from space, serving as a "storm patrol" for early warnings, an aid to weather analysis and forecasting, and a research tool for atmospheric scientists (Wexler and Caskey 1963). The images revealed surprising new cloud features: ocean storms, including the spiral band structure of hurricanes and an unreported typhoon near New Zealand; the unexpectedly great extent and structure of mountain wave clouds over South America; and rapid changes occurring during cyclogenesis. In a posthumous article published in 1965, Harry Wexler (Box 2.2, Figure 2.6) wrote, "The TIROS satellites disclosed the existence of storms in areas where few or no observations previously existed, revealed unsuspected structures of storms even in

BOX 2.1
James A. Van Allen (1914-2006)

James A. Van Allen (Figure 2.2) was born in Mount Pleasant, Iowa; he earned a Ph.D. in physics from the University of Iowa and spent almost his entire career there in the Department of Physics and Astronomy. During World War II he served in the Navy, working at the Carnegie Institution of Washington and the Johns Hopkins University Applied Physics Laboratory, where he helped develop and test the radio proximity fuse. After the war he developed instrument packages for upper-atmosphere and near-space scientific exploration and was head of development for the Aerobee sounding rocket.

In 1950, Van Allen hosted a dinner at his home where plans were initiated for the IGY—a coordinated, international, and comprehensive study of Earth conducted in 1957-1958. Two of the most prominent achievements of the IGY were the discovery of the Van Allen radiation belts and a new, pear-shaped model of Earth. Van Allen subsequently served as principal investigator for more than 25 space science missions. He was active in NASA, National Academy of Sciences, and American Geophysical Union affairs and was an articulate and outspoken advocate of small, inexpensive space missions.

FIGURE 2.2 James A. Van Allen, in his office on the University of Iowa campus in Iowa City, 1990. SOURCE: Photo by Tom Jorgensen, University of Iowa. Reprinted with permission from Tom Jorgensen, University of Iowa.

FIGURE 2.3 Explorer 1 with architects (left to right) William H. Pickering, director of the Jet Propulsion Laboratory; James A. Van Allen, chief scientist; and Wernher von Braun, leader of the Army's Redstone Arsenal team. SOURCE: Jet Propulsion Laboratory.

FIGURE 2.4 TIROS 1 and technician. SOURCE: National Archives Photo 370-MSP-4-147.

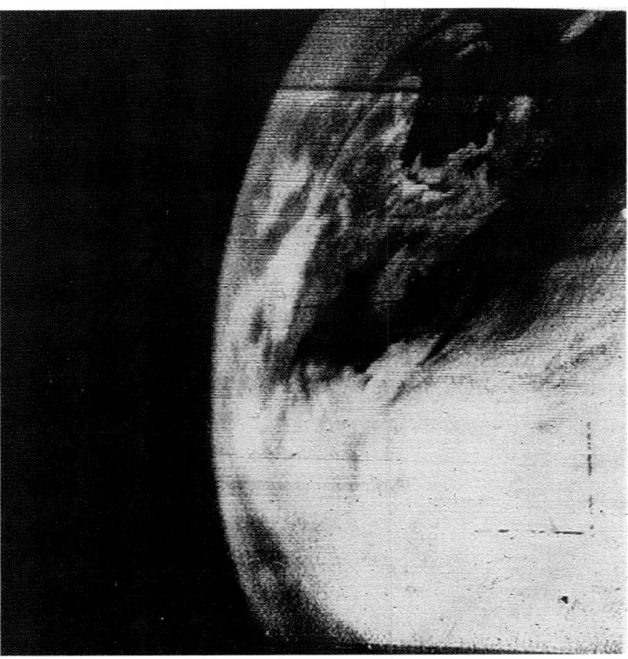

FIGURE 2.5 TIROS 1: first TV picture of weather from space, April 1, 1960. SOURCE: NASA.

areas of extensive observational coverage, depicted snow fields over land, ice floes over water, and temperature patterns on land and ocean as well as temperatures of tops of cloud layers."

Because its accomplishments were clearly accessible to the general public, the TIROS program enjoyed strong political support. The series of 10 TIROS satellites proved to be reliable and operationally successful, providing proof of the concept that sustained weather observations from space were possible. In 1966 this was continued as the Environmental Science Services Administration series, the Improved TIROS Operational System, TIROS M, TIROS N, and the National Oceanic and Atmospheric Administration (NOAA) series of satellites.

Launched by NASA on October 13, 1959, Explorer 7 carried a number of instruments including solar X-ray, Lyman-alpha, cosmic radiation, and micrometeor detectors. Significantly, it also carried University of Wisconsin, Madison scientist Verner Suomi's (Box 2.3, Figure 2.8) improved radiometer, which took the first Earth radiation measurements from space and initiated the era of satellite studies of climate. These observations established that Earth's energy budget varies markedly due to the effect of clouds and other absorbing constituents (Suomi 1961). Beginning in 1963, in collaboration with Robert Parent, Suomi developed a spin-stabilized camera that continuously

BOX 2.2
Harry Wexler (1911-1962)

Harry Wexler (Figure 2.6) was born in Fall River, Massachusetts, and earned a Ph.D. in meteorology from the Massachusetts Institute of Technology under C. G. Rossby. He spent most of his career with the U.S. Weather Bureau and also served as an instructor of military weather cadets during World War II. As head of research for the Weather Bureau, Wexler participated in the development of a number of new technologies, including airborne observations, sounding rockets, radar, the use of electronic computers for numerical weather prediction and general circulation studies, and satellite meteorology. He also served as chief scientist for the U.S. expedition to the Antarctic for the IGY and established a number of atmospheric baseline measurements of trace gases, including carbon dioxide and ozone. From the early 1950s, Wexler promoted the use of satellites in meteorology. Wexler, who played a central role in the development of Explorer and TIROS, foresaw that information gathered from satellites would be of great value for severe weather warnings, measurement of Earth's heat budget, and detection of environmental changes, both immediate and long term. Always interested in global studies of weather and climate, Wexler was an enthusiastic promoter of the World Weather Watch, which became a reality in 1963, shortly after his death.

FIGURE 2.6 Harry Wexler. SOURCE: Wexler and J. E. Caskey, Jr. (1963). North-Holland Publishing Co., Amsterdam, 1963.

monitored the weather and its motions over a large fraction of Earth's surface. Their "spin-scan camera" first flew on NASA's ATS 1, a spin-stabilized communications satellite launched into geostationary orbit in December 1966 (Suomi et al. 1971).

Suomi's "gadget," as he called all his inventions, produced spectacular full-disk images of Earth. It allowed scientists to observe weather systems as they developed, instead of glimpsing small bits at odd intervals, and it provided the first full-Earth disk: high-quality, cloud-cover pictures taken from a geostationary satellite. The image had a resolution of about 3 km and allowed meteorologists to identify and study significant cloud patterns, including those in the tropics, their change with time, the structure of storms, and the way synoptic-scale weather interacted with local phenomena. The spin-scan system, used in conjunction with ground and tropospheric balloon observations, also contributed to quantitative measurements of the dynamics of air motion, cloud heights, and the amount of atmospheric pollution. Using the communications capabilities of ATS 1, NASA was able to send cloud images and weather data to ground stations in the United States, Canada, Japan, Australia, and islands in the Pacific. The weather satellite images seen today on worldwide television and the Internet are direct descendents of Suomi's invention. They are essential for warning the public of tropical storm landfalls and other potential natural disasters.

A color version of the spin-scan camera flew on ATS 3 in 1967. It was used to detect the genesis of severe storms in the American Midwest, document the complete life cycle of hurricanes, and support the Barbados Oceanographic and Meteorological Experiment (BOMEX) study of energy exchanges between the ocean and the atmosphere. Subsequent experiments involved a Visible and Infrared Spin Scan Radiometer (VISSR) on Synchronous Meteorological Satellites (SMS) 1 and 2, launched in 1974 and 1975, and the 13 Geostationary Operational Environmental Satellites (GOES) launched since 1975. These experiments enabled Suomi and his team to document that Earth absorbed more of the Sun's energy than originally thought and to demonstrate that it was possible to measure and quantify seasonal changes in the global heat budget.

One of Suomi's important contributions to satellite data processing was the Man-Computer Interactive Data Access System, developed by an interdisciplinary team of electronics and computer engineers and programmers at the Space Science and Engineering Center at the University of Wisconsin, Madison. This data system provided an important interface between the user, the computer, the databases, and ultimately, real-time sensors, including satellite-based instruments, ground-based radar, and conventional and automated meteorological stations (Chatters and Suomi 1975). It was used for both research and operational purposes in support of data collected during the Global Atmospheric Research Program (GARP) Atlantic Tropical Experiment in 1974 and the First GARP Global Experiment, subsequently known as the Global Weather Experiment (GWE), 1977-1979.

The GWE, at the time the largest fully international scientific experiment ever undertaken, linked in situ and satellite data with computer modeling in an attempt to improve operational weather forecasting, determine the ultimate range of numerical weather prediction, and develop a scientific basis for climate modeling and prediction. In this experiment, worldwide surface and upper-air observations from satellites, ships, land stations, aircraft, and balloons were combined with global coverage provided by two U.S. GOES satellites, the European Meteosat, the Russian Geostationary Operational Meteorological Satellite (GOMS), and the Japanese Geostationary Meteorological Satellite (GMS) (Figure 2.7).

Launched into sun-synchronous polar orbit between 1964 and 1978, the Nimbus series of seven satellites contributed to our understanding of the atmosphere, land surface and ecosystems, weather, and oceanography and constituted the nation's "primary research and development platform for satellite remote-sensing of the Earth" (Figure 2.9). According to NASA, "Each mission taught scientists not only something new about the Earth system, but also something new about how to create, operate, and improve the technology for observing the Earth from space" (NASA Earth Observatory, *http://earthobservatory.nasa.gov/Study/Nimbus/*).

Nimbus 1 (1964) provided the first global images of clouds and large weather systems. Flying a medium-resolution infrared radiometer, Nimbus 2 (1966-1969) mapped the distribution of water vapor and carbon dioxide in the atmosphere, measured the temperature of the ocean, and clearly revealed the outlines of major ocean currents. Nimbus 3 (1969-1972) carried an advanced navigation and locator communications system, a forerunner of the Global Positioning System (GPS), that was used to track and interrogate neutral buoyancy balloons; it also opened up the possibility of vertical temperature and water vapor soundings and the ability to measure Earth's radiation budget above the atmosphere, allowing for estimates of zonal poleward heat transport. Nimbus 4 (1970-1980) flew spatially scanning infrared sounders and collected global observations of the ozone layer. Nimbus 5 (1972-1983) carried microwave and stratospheric sounders, measured rainfall over the oceans, and mapped and monitored sea ice. Nimbus 6 (1975-1983) improved measurements of atmospheric temperature at different altitudes.

The long-lived Nimbus 7 (1978-1994) carried the Coastal Zone Color Scanner (CZCS; see Chapter 8), which provided data until 1986; the Total Ozone Mapping Spectrometer (TOMS; see Chapter 5), which failed in 1993; and six other improved versions of sensors previously flown. The cavity radiometer aboard the Earth Radiation Budget Experiment of the Nimbus 7 satellite provided the first precise measurements of total solar irradiance reaching Earth. The technology and lessons learned from the Nimbus missions stand behind most of the Earth-observing satellites that NASA and NOAA have launched since 1978.

NASA's efforts in satellite geodesy date to 1962 when planning began for the National Geodetic Satellite Program, with the initial goal of developing "a unified world datum accurate to ±10 m and to refine the description of the earth's gravity field." A satellite-based laser tracking

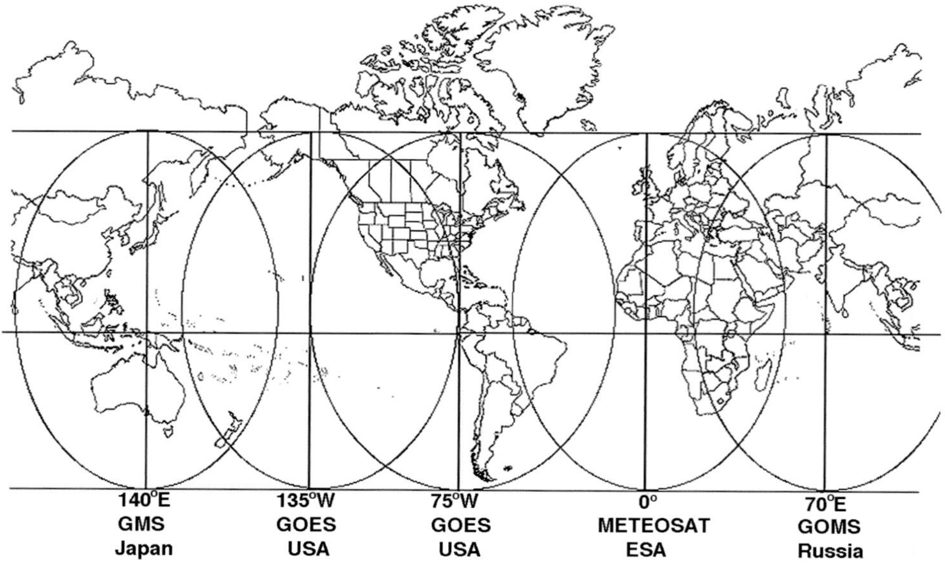

FIGURE 2.7 During the GWE, five international geostationary satellites supported global observations of cloud-tracked winds. SOURCE: NOAA (1984).

BOX 2.3
Verner Edward Suomi (1915-1995)

Verner Edward Suomi (Figure 2.8) was born in Eveleth, Minnesota, received a Ph.D. in meteorology from the University of Chicago and spent most of his career at the University of Wisconsin, Madison, where he was co-founder of the Space Science and Engineering Center, a premier institution dedicated to atmospheric research and instrument development for satellites and other spacecraft.

Suomi is considered one of the founding fathers of satellite meteorology. His innovations in scientific instrumentation, data processing, and analysis have substantially improved our understanding of weather and climate. Suomi's professional career combined inventiveness with a keen ability to mobilize human and financial resources in support of his ideas and projects.

FIGURE 2.8 Verner Edward Suomi. SOURCE: University of Wisconsin-Madison. Reprinted with permission by the University of Wisconsin, Madison.

system immediately returned results of ±1 m, which was an order-of-magnitude improvement, with an additional order-of-magnitude improvement possible (NASA 1970).

In 1978, based on its experience with planetary probes, JPL launched Seasat, an experimental satellite carrying a variety of oceanographic sensors, including imaging synthetic aperture radar, altimeters, radiometers, and scatterometers. The instruments measured ocean surface topography, boundary-layer ocean wind speed and direction, sea surface temperature, and polar sea ice conditions. Although Seasat collected data for only 105 days, it inspired the future use of imaging radars on NASA's Space Shuttle, Topography Experiment (TOPEX)/Poseidon, and a number of satellites flown by the Europeans, Canadians, and Japanese. These satellite-borne radars are able to detect minute changes in surface features due to tectonic, volcanic, hydrologic, or anthropogenic activity.

INSTRUMENT AND TECHNOLOGY DEVELOPMENT

Although it carried no remote sensing instruments, the orbital decay of the first Earth satellite, Sputnik 1, provided information about the density and dangers of the near-space environment. The operation of its two radio transmitters provided clues regarding the electron density of the ionosphere and indicated that the satellite's pressurized nitrogen compartment had not been punctured by micrometeorites (Hagfors and Schlegel 2001).

Many observations of the Earth system from space have been conducted using optical sensors. TIROS 1 housed two television cameras and stored the information on magnetic tapes for later transmission to ground stations. TIROS 2 contained an infrared sensor in addition to the two cameras and a new attitude control system using Earth's magnetic field. Subsequently, optical sensors were adapted for many different applications in all disciplines of the Earth sciences. Video cameras soon gave way to medium- and high-resolution radiometers in all wavelengths. TIROS 6, launched in 1962, made the first measurements of snow cover from space using infrared sensors. This technology is also used to observe sea ice coverage and the temperature of cloud tops and the sea surface.

Since the atmosphere is virtually transparent to microwave radiation, these sensors can penetrate clouds to make ground measurements in all weather conditions. The first such passive microwave remote sensing system for satellites was launched on the Russian Cosmos 243 (1968) and Cosmos 384 (1970) (Johannessen et al. 2001). The Electrically Scanning Microwave Radiometer (ESMR) was flown on Nimbus 5 (1972-1983) over the Arctic to detect sea ice coverage, where cloud cover frequently interfered with infrared technology. Subsequently, this technology was further developed, and the Scanning Multichannel Microwave Radiometer (SMMR) on Nimbus 7 (1978-1994), together with the Defense Meteorological Satellite Program (DMSP) Special

Sensor Microwave/Imager (SSM/I), have provided the longest and most regular time series of global sea ice data at a resolution of typically 25 × 25 km (Johannessen et al. 2001). Moreover, the microwave radiation emitted by atmospheric oxygen and water vapor were used for vertical soundings, especially in the upper atmosphere.

In 1972, ERTS 1 (Earth Resources Technology Satellite), later renamed Landsat 1, provided new land-based applications for optical sensors. Landsat carried a return beam videocon (RBV) and a multispectral scanner (MSS) that imaged Earth from an altitude of 900 km with green, red, and two infrared spectral bands at 80-m resolution. Since 1972, Landsats have provided the longest, continuous global record of land cover and its historical changes in existence. Landsat is the premier technology supporting the new geographical field of land-cover science, part of Earth system science.

In the 1970s, laser technology was first employed in combination with satellites (Laser Geodynamics Satellites [LAGEOS] 1 and 2) that were designed for maximum reflectivity to allow for study of Earth's geoid and the movements of tectonic plates. Interestingly, this type of satellite does not contain any instrumentation, and for that reason the first such satellite launched in 1976 is still operational today.[1] Spaceborne synthetic aperture radar enables observation of sea ice with much better accuracy than visible and passive microwave methods, as proven by Seasat, the European Remote Sensing Satellite (ERS 1), Canada's RADARSAT, and Europe's Envisat (Figure 2.10) (Johannessen et al. 2001). Although Seasat was able to collect data only for 105 days, it pioneered the exploitation of radar technology and the microwave range to measure ocean topography and winds. Its success led to important follow-on missions such as TOPEX/Poseidon and QuikScat.

FIGURE 2.9 Artist's drawing of the general design of the Nimbus series of satellites. SOURCE: C. R. Madrid, ed. (1978). The Nimbus 7 Users' Guide, Goddard Space Flight Center, NASA.

CONCLUSION

In the first two decades of the space age, six nations designed and launched Earth-orbiting satellites for scientific purposes: the Soviet Union (1957), the United States (1958), France (1965), Japan (1970), China (1970), and the United Kingdom (1971). The European Space Agency, a consortium of 17 member nations, was founded in 1974. International cooperation has been an important aspect of the satellite legacy: the Global Weather Experiment (1979) demonstrated what was possible in observation, modeling, understanding, and prediction. Today, such experiments are conducted every

[1]The projected lifetime of the LAGEOS satellites is more than 200,000 years.

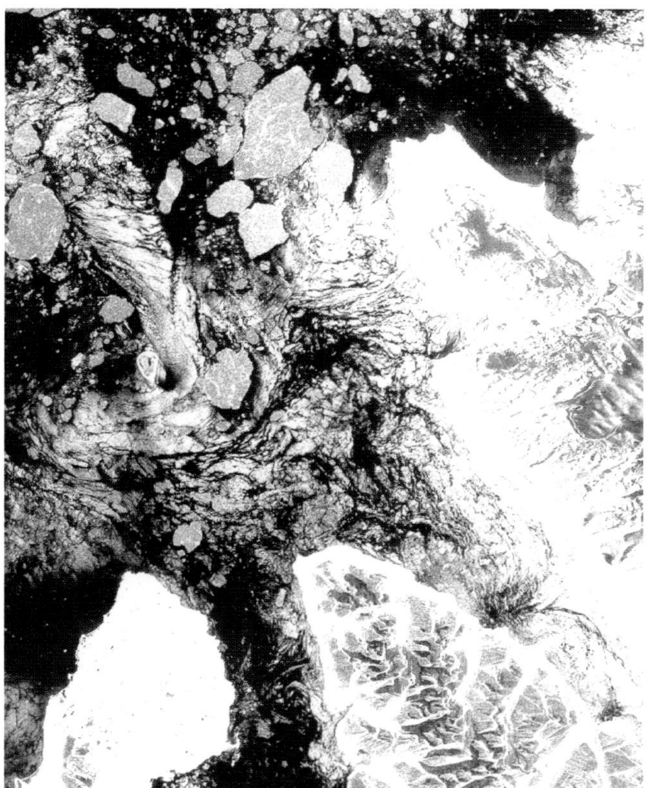

FIGURE 2.10 Detail of sea ice off the west coast of Greenland from ERS 1. SOURCE: ERS-1 User Handbook, SP-1148, European Space Agency, Paris. Reprinted with permission by the European Space Agency.

day—not just for the weather but to explore and monitor all components of the Earth system. International cooperation has other dimensions as well. When the U.S. GOES-W satellite failed in 1989, it was replaced by a French Meteosat. From 2003 to 2005, GOES 9 was on loan to Japan to cover a significant gap in Japan's meteorological satellite coverage.

Our imperfect appreciation of how much has been accomplished in the past 50 years may well reflect a gap in historical comprehension. The "missile gap" widely feared in 1957 was not real, and the militarization of space, fortunately, has not happened yet. International cooperation, rather than competition, has become the dominant theme in the space age, not only in satellite hardware but also in creative analysis and use of satellite data. Just what would the world be like without scientific satellite remote sensing and services? It would be much like 1957. The Moon would be Earth's only satellite. There would be no weather eyes in the sky—no global ability to monitor changes in atmospheric composition, in ecosystems and land use, in climate variability and change, in Earth's surface and land-use changes, in ocean physical and biological processes, or in ice sheets.

The complexity of today's bureaucratic and budgetary practices has created a time delay problem. Many of the expert consultants who made presentations to this committee mentioned the ease of moving among agencies and programs in the early days and how new ideas were rapidly tested and research results found quick expression in operational systems. The ideas generated by Wexler, Van Allen, Suomi, and others were well supported and well funded, with comparative ease. The ferment of ideas was supported by rapid development and short time lags between concept and launch. This no longer seems to be the case.

A final issue derives from findings of the National Academies' recent decadal survey of Earth science missions that reports an actual "satellite gap" in which space resources will decrease dramatically compared to the scientific challenges associated with, for example, climate change research. These satellite data gaps also stand in stark contrast to the stunning and growing needs for space-based information by the world's inhabitants (NRC 2007a).

At the dawn of the space age, the very first satellites provided new and important scientific knowledge of the Earth system that could not be obtained by any other means. The first two decades were extremely exciting, yet new discoveries, transformative breakthroughs, proof of concepts, improved understanding, and societal benefits have continued to accumulate, as the following chapters in this report document. How can we compare the cumulative amount spent by the nations of the world on the scientific study of Earth from space with the inestimable value of understanding our home planet?

3

Weather

Because many aspects of daily life are affected by the weather, understanding and predicting the weather has been a human quest for millennia. Space-based observations have played pivotal roles in the history of weather forecasting. Their contributions to forecasting at all spatial scales can be grouped into three areas, which are described in this chapter: weather imagery, atmospheric properties, and numerical weather forecasting.

Fundamentally, weather forecasting is a four-dimensional problem, involving three spatial dimensions and time. The air that is now affecting point B was yesterday at point A and tomorrow will be at point C. Similarly, the storm system centered at point A yesterday is centered at point B today and has a different shape, size, and intensity than it did yesterday. To forecast tomorrow's weather at a point, one must know today's weather over a broad region surrounding that point. The farther one wants to forecast into the future, the larger the area must be where one knows the weather today.

In 1846 the state of the art in weather forecasting was succinctly stated by François Arago[1]: "Whatever may be the progress of sciences, NEVER will observers who are trustworthy, and careful of their reputation, venture to foretell the state of the weather." In the United States, however, isolated observers were communicating among themselves to understand the horizontal extent of the weather, but they communicated by mail, which meant that the weather could not be forecasted, only understood in retrospect.[2] With the development of the telegraph, Joseph Henry[3] and James Espy[4] experimented with transmitting weather observations in real time to a central site, where weather maps could be drawn and forecasts made. By 1870, President Ulysses S. Grant signed a joint resolution of Congress authorizing the Secretary of War to establish a "Division of Telegrams and Reports for the Benefit of Commerce" as part of the U.S. Army Signal Service Corp (NWS 2006). This division became the Weather Bureau in 1890 and the National Weather Service in 1967.

The development of radiosondes (weather balloons that transmit information to a fixed location) in the 1930s yielded measurements that resulted in major advances in weather forecasting by adding the crucial vertical dimension to meteorological observations. About half of the stations in the Integrated Global Radiosonde Archive depicted in Figure 3.1 make observations twice per day. Though numerous, these stations cannot give the desired global picture of the current state of the weather. The ocean is especially data sparse.

After World War II, rockets were sufficiently advanced to be able to lift cameras high above the clouds to take photographs of weather systems and to indicate the potential for weather observations by Earth-orbiting satellites. This led Harry Wexler[5] to publish a paper in 1954—3 years before the launch of the first satellite and 6 years before the Television Infrared Observation Satellite (TIROS) (see Chapter 2)—titled "Observing the Weather from a Satellite Vehicle" (Wexler 1954). Thus, weather observations from space were not serendipitous but were eagerly anticipated.

Since the beginning of the space age, perhaps 200 weather satellites have been launched, as nations around the world recognized their value and as technology advanced to make more capable instruments possible.

WEATHER IMAGERY

The first weather satellites attempted simply to "take pictures" from space. By the mid-1960s, engineers had

[1]1786-1853, director of the Paris Observatory and permanent secretary of the French Academy of Sciences.

[2]For a history of these developments, see Fleming (1990).

[3]1797-1878, first secretary of the Smithsonian Institution and one of the founding members of the National Academy of Sciences.

[4]1785-1860, America's first national meteorologist.

[5]1911-1962, chief of the Scientific Services Division of the Weather Bureau.

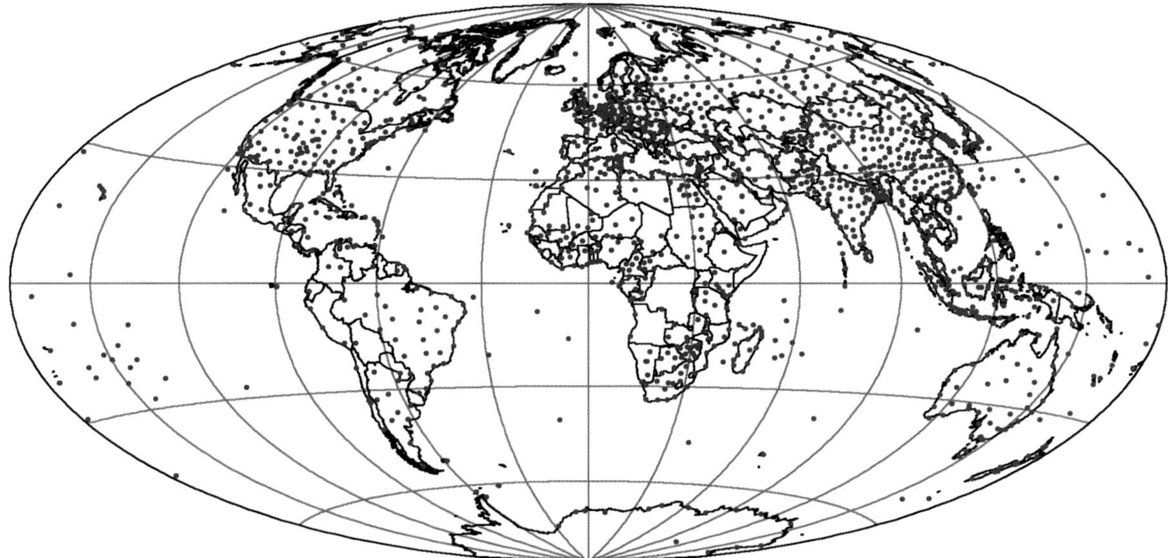

FIGURE 3.1 Locations of Integrated Global Radiosonde Archive stations. SOURCE: Data from the National Climatic Data Center, Durre et al. (2006). Reprinted with permission from the American Meteorological Society, copyright 2006.

developed the capability to fly satellites in sun-synchronous orbits,[6] in which an instrument on a single satellite could view the entire Earth twice per day, once in daylight and once at night (Figure 3.2). Then meteorologists could tile the pictures together to form the long-sought global picture of Earth's weather (Figure 3.3).

Also during the mid-1960s, the first geostationary satellites were launched. These satellites orbit Earth in the equatorial plane at the same angular velocity that Earth rotates on its axis; thus, they stay "stationary" over the same point on the equator. Although they do not view the entire Earth but only one hemisphere (Figure 3.4), they can make images frequently, not just twice per day. These images can be assembled into movies that allow forecasters to watch the weather in motion. This is an invaluable tool for weather analysis and forecasting (Box 3.1). Geostationary satellites rapidly became the choice of weather services worldwide, such that today they form a ring around the equator, providing coverage of the entire tropics and midlatitudes.

Many accomplishments in weather forecasting have been achieved using the imagery from weather satellites. Only a few can be mentioned in this document. Perhaps the most dramatic accomplishments relate to observing and predicting hurricanes and tropical storms. In 1900 a "surprise" hurricane roared out of the Gulf of Mexico over Galveston Island killing at least 8,000 people; it was the largest natural disaster in the United States (Blake et al. 2006). Since then an important scientific accomplishment occurred sometime in the 1960s: with the continuous monitoring of weather by satellites, no tropical cyclone anywhere on Earth escapes detection (Figure 3.5). Indeed, Robert C. Sheets, former director of the National Hurricane Center (NHC), has written:

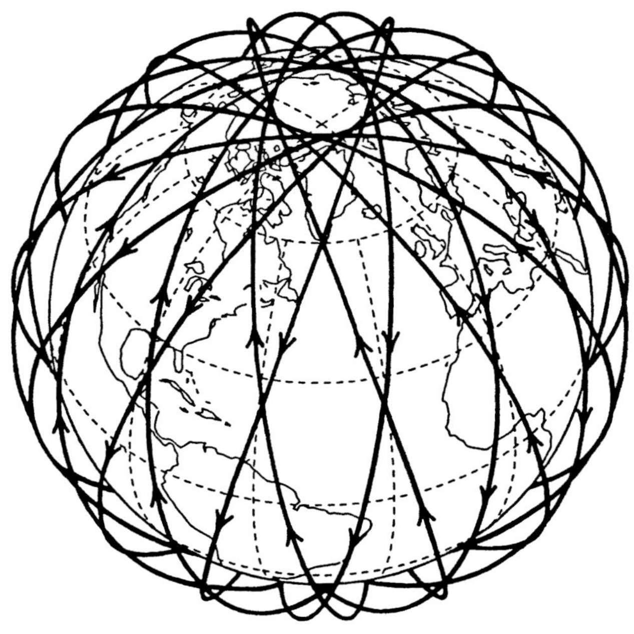

FIGURE 3.2 One day's orbits of a sun-synchronous satellite. A single instrument views the entire Earth. SOURCE: Kidder and Vonder Haar (1995). Copyright Elsevier, 1995.

[6]A circular, near-polar orbit in which the orbital plane keeps a constant relationship to the Sun-Earth line, such that the satellite passes over a point on the Earth near the same time every day.

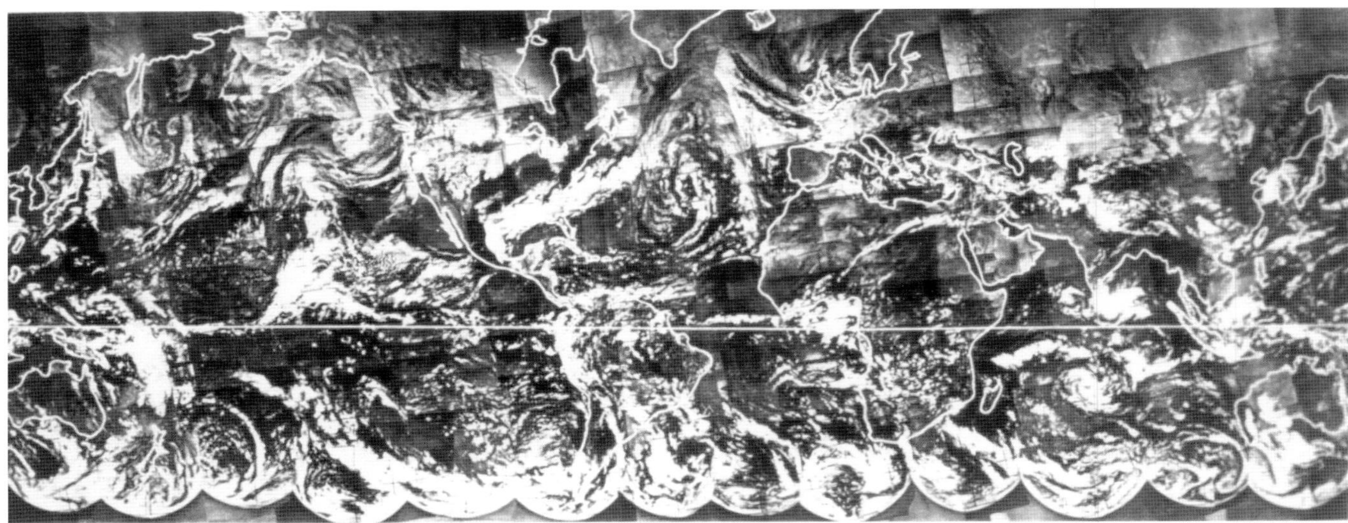

FIGURE 3.3 First complete view of the world's weather, photographed by TIROS 9, February 13, 1965. Image assembled from 450 individual photographs. SOURCE: Publication of the National Oceanic and Atmospheric Administration (NOAA), NOAA Central Library.

FIGURE 3.4 Example of the hemispheric coverage of a geostationary satellite. Taken by NASA's Applications Technology Satellite 3 (ATS 3) at 1402 UTC on July 21, 1970. Note that Tropical Storm Becky can be seen in the Gulf of Mexico near Florida. SOURCE: NOAA Photo Library.

BOX 3.1
Weather "Movies"

One of the best ways to understand the impact that observations from space have had on weather forecasting is to watch the weather in motion in a sequence of satellite images. Unfortunately, this is impossible in a printed document. Readers are urged to visit one of the many Internet sites that offer satellite movies and try a bit of "nowcasting": Where is that storm going, and when will it arrive at the reader's location? Here are a few sites:

- NOAA's Geostationary Operational Environmental Satellites (GOES) site: *http://www.goes.noaa.gov* (click on the MPEG or Java Applet icon)
- Japan Meteorological Agency's Multi-functional Transport Satellite (MTSAT) site: *http://www.jma.go.jp/en/gms/*
- EUMETSAT's Meteosat site: *http://www.eumetsat.int* (under "Image Gallery" choose "Real-time Images")

> The greatest single advancement in observing tools for tropical meteorology was unquestionably the advent of the geosynchronous satellite. If there was a choice of only one observing tool for use in meeting the responsibilities of the NHC, the author would clearly choose the geosynchronous satellite. (Sheets 1990)

From these observations we learned that tropical cyclones go through a life cycle that can be recognized and categorized in satellite imagery. The Dvorak scheme (Dvorak 1975) for estimating the intensity of tropical cyclones, which ranks storms between 1 and 8 based on wind speed and other features, is used worldwide. Tropical cyclone studies have benefited tremendously from satellite data (e.g., Special Sensor Microwave/Images, Advanced Microwave Sounding Unit, and QuickScat), which have been used to develop algorithms for monitoring and predicting hurricane intensity, tracks, and wind structures (Kidder et al. 1980, Demuth et al. 2000).

Since the advent of satellite imagery, scientists have learned to remotely identify many previously known weather features: fronts, high- and low-pressure systems, fog, low clouds, cirrus, and thunderstorms (Bader et al. 1995). Satellite observations also led to the discovery of thunderstorm clusters, called mesoscale convective complexes, which are unrelated to classical storm systems (Maddox 1980). Scientists also learned that rain-cooled air from thunderstorms descends and spreads out at the ground, producing an outflow boundary. When this boundary interacts with an adjacent storm, the intensity and destructive potential of both storms is amplified (Purdom 1976, 1986). In addition, satellite images revealed that a cold, V-shaped structure in the anvil of a thunderstorm is a signature of severe weather (Fujita 1978). It was also discovered that atmospheric turbulence is signaled by cloud patterns in the lee of mountain ranges (Ellrod 1989). Data from weather satellites allowed scientists to identify and forecast many other atmospheric phenomena too numerous to mention here (Kidder and Vonder Haar 1995).

ATMOSPHERIC PROPERTIES

In addition to taking pictures from space, satellites have allowed scientists to make radiometric measurements of the electromagnetic spectrum, from the ultraviolet to the microwave regions. From these measurements, scientists are able to retrieve properties of the atmosphere that are important to forecasters, especially the vertical temperature structure, winds, and moisture content (see Chapters 4 and 5), which are essential for numerical weather prediction.

Temperature profiles can be retrieved by several means. Given measurements at several wavelengths near an absorption band of a well-mixed gas, such as the 15-μm band of carbon dioxide or the 5-mm band of oxygen, the radiative transfer equation can be used to retrieve a temperature versus height profile that is consistent with the measured radiances (Chahine 1968, Smith 1970; see also Box 5.2). This has been done with a large number of satellite instruments starting with the Satellite Infrared Spectrometer (SIRS) and the Infrared Interferometer Spectrometer (IRIS), both launched on the Nimbus 3 satellite on April 14, 1969. Recently, the Atmospheric Infrared Sounder (AIRS) on the Aqua satellite (launched May 4, 2002) has provided high vertical resolution soundings using an infrared instrument that measures atmospheric radiation at 2,378 wavelengths (Chahine et al. 2006).

Radio occultation provides another way to measure atmospheric temperature profiles. Radio signals from Global Positioning System (GPS) satellites are refracted (bent) as they pass through the atmosphere. This bending angle can be measured from a second satellite. A sequence of refraction angles are measured as the GPS satellite rises or sets through the atmosphere. These measurements can be converted into a vertical profile of index of refraction of the atmosphere and thus into a vertical temperature sounding with high vertical resolution (Ware et al. 1996). The first such instrument, GPS/MET, was launched on the MicroLab 1 satellite on April 3, 1995; a constellation of six satellites (Formosa Satellite [FORMOSAT-3]/Constellation Observing System for Meteorology, Ionosphere, and Climate [COSMIC]) was launched on April 14, 2007.

A second property of the atmosphere that is necessary for weather forecasting is its water vapor content. It can be

FIGURE 3.5 At 1745 UTC on August 28, 2005, Hurricane Katrina was observed by the Geostationary Operational Environmental Satellite (GOES 12) near the time of its maximum wind speed, 150 knots (173 miles per hour). SOURCE: National Hurricane Center. Reprinted with permission from the National Hurricane Center, copyright 2005.

retrieved from satellite measurements by some of the same methods that are used to retrieve atmospheric temperature (e.g., Weng et al. 2003; see also Box 5.2). Figure 3.6 is an example of the vertically integrated water vapor content of the atmosphere over the ocean measured through clouds in the microwave portion of the spectrum. These images are used by forecasters to monitor tongues of moisture from the tropical oceans that can cause heavy rain and flooding when they encounter land.

Winds, or the atmospheric flow field, must be known to forecast the weather. Scatterometers, such as QuikScat (launched June 20, 1999), which measure wind speed and direction near the ocean surface and have "revolutionized the analysis and short-term forecasting of winds over the oceans at NOAA's Ocean Prediction Center" (Von Ahn et al. 2006), are discussed in Chapter 8. Another way to measure winds is to track clouds in sequences of satellite imagery, usually geostationary imagery. Small clouds travel with the wind. By observing the location of the same cloud in two successive satellite images and knowing the time difference between the images, the wind speed and direction can be calculated (Hubert and Whitney 1971). Figure 3.7 shows an example of the winds obtained by tracking clouds.

Finally, there are several other atmospheric parameters retrieved from satellite data that are useful to forecasters and are beginning to be used in numerical weather prediction models. For example, wind speeds around tropical cyclones are important for mariners and emergency managers. They are estimated by the Dvorak technique (mentioned above) and also by using microwave soundings. Microwaves penetrate the clouds and allow measurement of the magnitude of the warm core of tropical storms and its radial gradient, from which wind speeds can be derived (Kidder et al. 1980, Demuth et al. 2004).

Precipitation is a fundamental part of the hydrologic cycle and is further discussed in Chapter 6; it is often one of the first concerns when people think about the weather. Many satellite techniques have been developed to estimate rainfall (see, e.g., Barrett and Martin 1981). Today, daily rainfall estimates are available worldwide on the Internet. The most advanced precipitation estimation is from the Tropical Rainfall Measuring Mission (TRMM) satellite, launched on November 27, 1997 (Simpson et al. 1988, 1996). A joint U.S.-Japan mission, TRMM carries passive sensors of the visible to the microwave portion of the spectrum and is the first precipitation radar in space. Many other satellite-derived parameters are important to forecasters, including cloud height, cloud top temperature, and cloud phase; fog, smoke, and aerosol identification; and skin surface temperature (see, e.g., Kidder and Vonder Haar 1995).

NUMERICAL WEATHER PREDICTION

There are several reasons why Arago was wrong in 1846 and why today weather can be forecasted as much as 10 days ahead:

- The discovery of the mathematical principles that govern atmospheric flow and the change of phase of water,
- The invention of computers and the numerical techniques with which to solve these equations, and
- The development of observing systems to supply the needed initial state of the atmosphere.

Without doubt, one of the chief reasons for the success of weather forecasting is that Earth-orbiting satellites provide an accurate global initial atmospheric state, which the numerical weather prediction models project into the future. Today, satellite data constitute the vast majority of the data available for the initialization of numerical weather prediction models and has the greatest impact of any measuring technology in improving forecast skill. Table 3.1 lists the satellite data currently used to initialize models run by NOAA's National Centers for Environmental Prediction (NCEP).

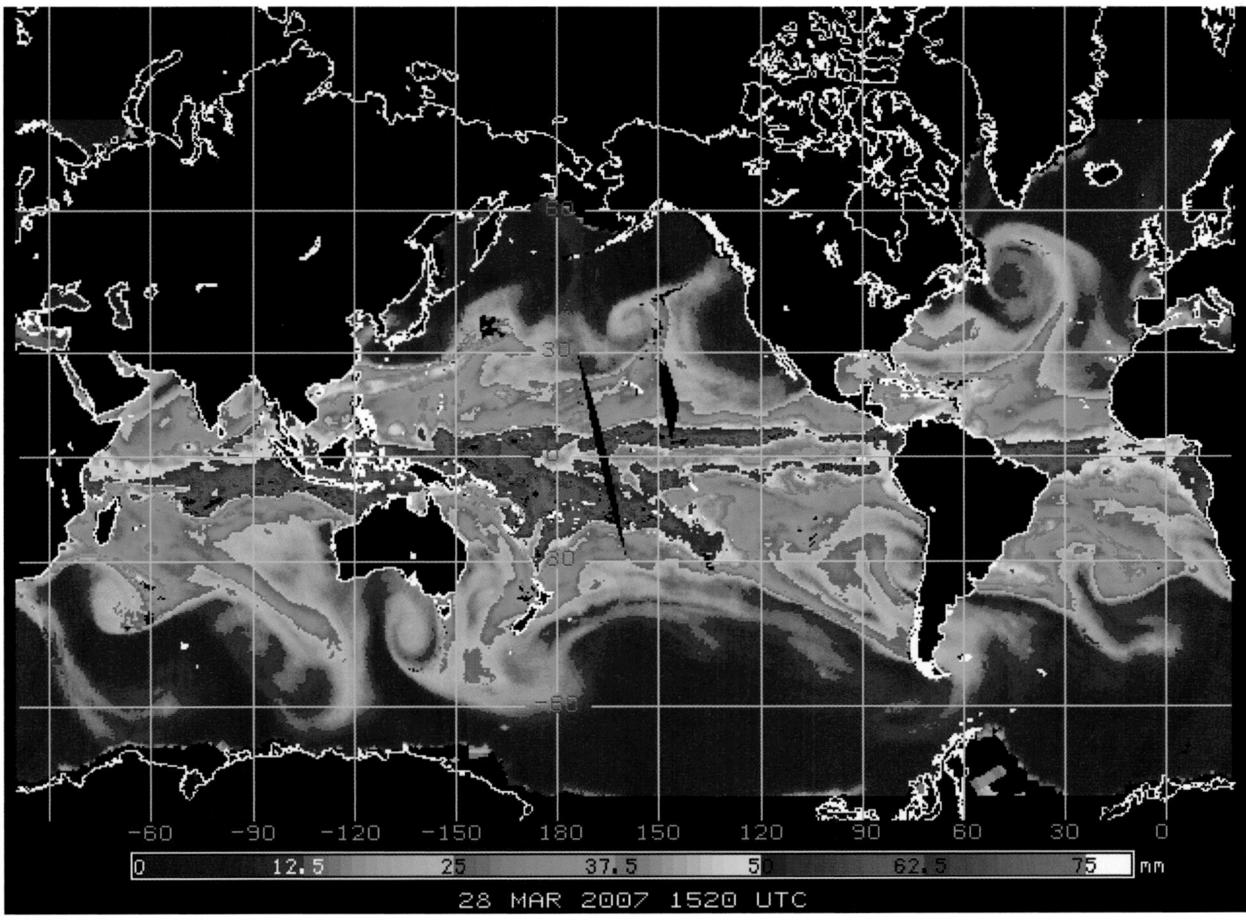

FIGURE 3.6 Vertically integrated water content of the atmosphere (in kilograms per square meter) derived from microwave measurements on six sun-synchronous satellites, three NOAA satellites, and three Defense Meteorological Satellite Program (DMSP) satellites. SOURCE: Data from NOAA; drawing by S. Kidder.

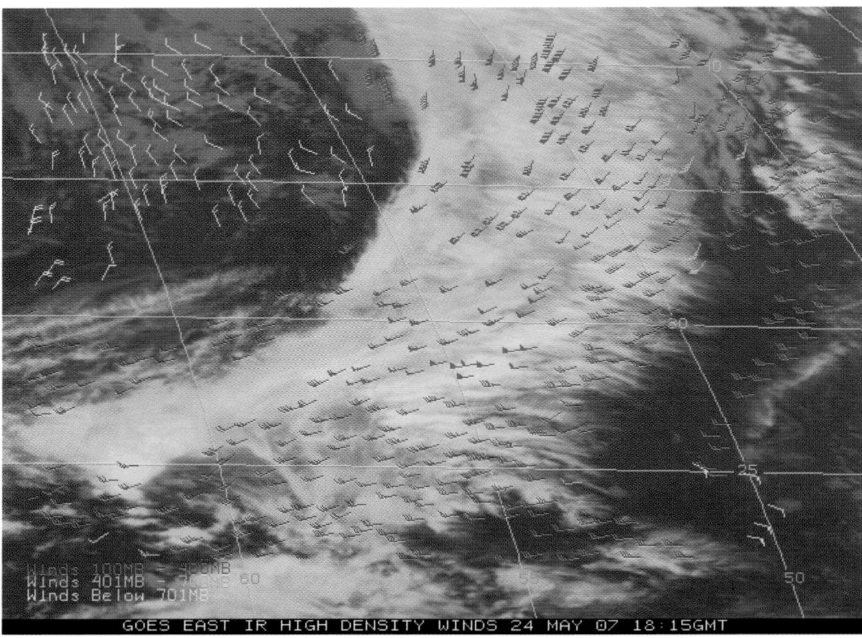

FIGURE 3.7 Winds obtained by tracking clouds in successive infrared images. The height of the cloud is determined by the cloud's temperature. Note that where there are no trackable clouds, no winds can be retrieved. SOURCE: NOAA.

TABLE 3.1 Satellite Data Used to Initialize Numerical Weather Prediction Models in 2006

Satellite Data
HIRS sounder radiances
AMSU-A sounder radiances
AMSU-B sounder radiances
GOES sounder radiances
GOES, Meteosat, GMS winds
GOES precipitation rate
SSM/I precipitation rates
TRMM precipitation rates
SSM/I ocean surface wind speeds
ERS-2 ocean surface wind vectors
QuikScat ocean surface wind vectors
AVHRR SST
AVHRR vegetation fraction
AVHRR surface type
Multisatellite snow cover
Multisatellite sea ice
SBUV/2 ozone profile and total ozone
AIRS
MODIS winds
Altimeter sea-level observations

NOTE: Monthly statistics on the data used in NCEP's models are available at *http://www.nco.ncep.noaa.gov/sib/counts/*.
HIRS = High-Resolution Infrared Radiation Sounder; AMSU = Advanced Microwave Sounding Unit; GOES = Geostationary Operational Environmental Satellites; SSM = Special Sensor Microwave; ERS = European Remote Sensing Satellite; AVHRR = Advanced Very High Resolution Radiometer; SST = sea surface temperature; SBUV = Solar Backscattered Ultraviolet.
SOURCE: Lord (2006).

Figure 3.8 shows a time series of one measure of the skill of a representative numerical weather prediction model for 3-, 5-, 7-, and 10-day forecasts. The top line for each set of curves is for the northern hemisphere, where nonsatellite observations are plentiful; the bottom line is for the southern hemisphere, where nonsatellite observations are woefully few. Due largely to our increasing ability to use satellite observations effectively—that is, to assimilate the observations into numerical weather prediction models (e.g., Kalnay 2003)—the difference between northern hemisphere forecasts and southern hemisphere forecasts has steadily decreased, and the overall forecast skill has increased to the point that global 7-day forecasts are now as good as northern hemisphere 5-day forecasts were 25 years ago.

In addition, tests at NCEP show that data from just one satellite instrument, the Advanced Microwave Sounding Unit (AMSU), extend forecast usefulness by 1 day in the southern hemisphere and by about a half day in the data-rich northern hemisphere (Lord 2006). AIRS is also improving forecasting skill (Chahine et al. 2006), and there is evidence that satellite data, particularly QuikScat winds, are improving hurricane track forecasts (Zapotocny et al. 2007). Without question, improvement in numerical weather prediction—on which all forecasts more than a few hours ahead are based—is a major scientific accomplishment of Earth observations from space.

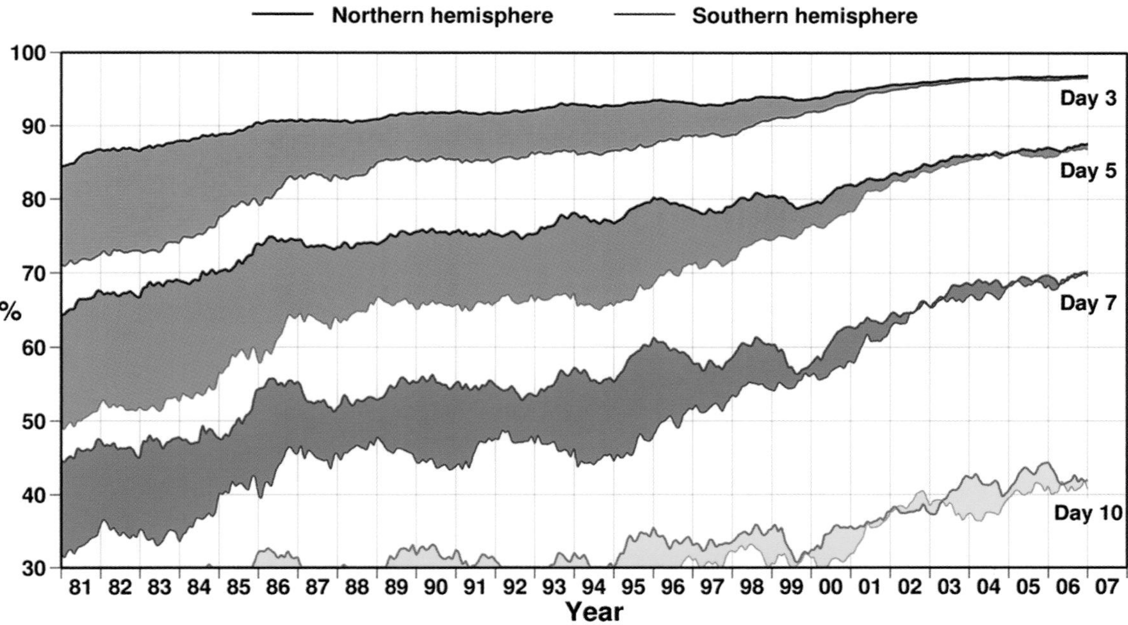

FIGURE 3.8 Anomaly correlation of 500 hPa height forecasts by the European Centre for Medium Range Forecasting. SOURCE: Updated from Simmons and Hollingsworth (2002). Reprinted with permission from the Royal Meteorological Society, copyright 2002.

4

Earth's Radiation Budget and the Role of Clouds and Aerosols in the Climate System

Earth orbit is the ideal location from which to measure the exchange of energy between Earth and space and its variability around the globe. The surface temperature of Earth is in energy balance when the solar radiation absorbed by Earth is balanced by the emission of thermal energy from Earth to space (Figure 4.1). If Earth gains energy from space, the surface temperature will warm up until the energy exchange with space is again balanced. This chapter illustrates how satellite measurements of solar energy output, Earth radiation fluxes, clouds, water vapor, and aerosols have improved our understanding of the climate system and its sensitivity to changes in atmospheric composition.

Solar energy emission is mostly in visible and near-infrared wavelengths, while Earth's emission is in thermal infrared wavelengths. Basic starting points for understanding the radiation balance are measurements of the energy coming from the Sun, the reflection and absorption of solar energy by Earth, and the export of energy from Earth by the emission of thermal infrared radiation to space. The greenhouse effect of the atmosphere is important in the energy balance and is driven largely by water vapor, clouds, and carbon dioxide. Noncloud aerosols are also very important in the climate system.

EARTH'S RADIATION BUDGET

Measurement of Earth's radiation budget was one of the earliest proposals for a scientific application of Earth-orbiting satellites on Explorer 7 (see Chapter 2 and House et al. 1986). Early measurements showed that Earth was a warmer and darker planet compared to presatellite estimates indicating that a greater poleward energy transport by the atmosphere and ocean was required (Vonder Haar and Suomi 1969, 1971). The quality of measurement has steadily increased since those early days. Earth-orbiting satellites now allow precise global measurement of Earth's thermal emission, the solar radiation reflected from Earth, and the energy coming from the Sun (Box 4.1, Figure 4.2). Monitoring of variability and change has become an increasingly important goal because Earth's climate is likely changing in response to human activities.

Accurate observations of the radiation balance (Figure 4.3) as a function of latitude allows direct measurement of the annual mean poleward transport of energy. Earth gains energy from space in the tropics and returns this energy to space at high latitudes. The poleward heat transport in the atmosphere and ocean warms the poles and cools the tropics and also plays a key role in determining the response of global climate to greenhouse gases. If atmospheric data are used to compute the atmospheric heat transport, oceanic heat transports can be inferred by subtracting atmospheric transport from the measured total transport. A measurement of the total required poleward energy flux from space provides independent data that can be used to test estimates of atmospheric and ocean heat fluxes based on in situ measurements. Estimates of oceanic heat fluxes from direct measurements of ocean current and temperature are difficult. Measurements from space provided the first estimates of poleward heat flux in the ocean, which is nearly as large as the atmospheric flux but reaches a maximum at a relatively low latitude of about 20 degrees, while the atmospheric flux peaks at about 50 degrees latitude (Vonder Haar and Oort 1973, Trenberth et al. 2001).

Measurements from the Earth Radiation Budget Experiment (ERBE; Barkstrom et al. 1984) had sufficient accuracy and spatiotemporal resolution to allow the inference of the clear-sky radiation balance and thus measure the effect of clouds on Earth's radiation balance (Figures 4.4 and 4.5). This showed that clouds double Earth's albedo from 0.15 to 0.3 and reduce the emitted thermal radiation by 30 W/m^2 (Ramanathan et al. 1989, Harrison et al. 1990). These basic measurements provide a standard against which to test climate models.

The amount by which the atmosphere reduces the loss of thermal energy to space is one way to measure the strength

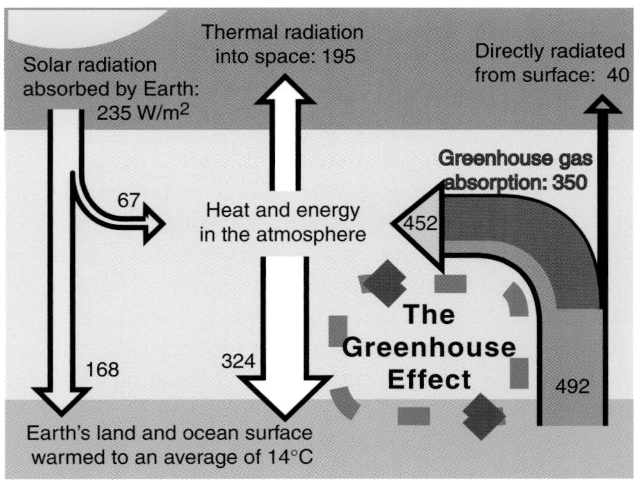

FIGURE 4.1 Simplified diagram of Earth's radiation budget. Energy is exchanged between three main sources: space (gray), the atmosphere (blue), and the surface (brown), expressed in watts per square meter (W/m^2) and derived from Kiehl and Trenberth (1997). Short-wavelength solar radiation (yellow arrow) enters the atmosphere and reaches land. A fraction is reflected back into space by the atmosphere or the surface (brown arrow). Another fraction of the energy is absorbed by the atmosphere and reemitted as long-wavelength radiation (white arrow) into space or back to the surface. Adding greenhouse gases increases the fraction of the energy absorbed by the atmosphere and reemitted back to the surface, which increases the surface temperature to balance the radiation budget between the atmosphere and surface. Under stable conditions, the total amount of energy entering the system from solar radiation will exactly balance the amount being radiated into space, thus allowing Earth to maintain a constant average temperature over time. Recent measurements indicate that the Earth is presently absorbing 0.85 ± 0.15 W/m^2 more than it emits into space (Hansen et al. 2005). SOURCE: Data from Kiehl and Trenberth (1997) and Hansen et al. (2005). Drawing by R. A. Rohde, University of California, Berkeley. Robert A. Rohde/Global Warming Art.

of the greenhouse effect. With measurements of the outgoing longwave radiation and observations of the surface temperature and emissivity, the greenhouse effect of the atmosphere at any location can be computed. The average strength of Earth's greenhouse effect is about 155 W/m^{-2}, but it varies from about 270 W/m^{-2} in moist, cloudy regions of the tropics to about 100 W/m^{-2} at high latitudes. The role of water vapor in the greenhouse effect has also been measured in this way (Raval and Ramanathan 1989, Rind et al. 1999, Inamdar and Ramanathan 1998). Global satellite measurements of water vapor using infrared sounding and microwave imaging data allowed isolation of the water vapor contributions to the greenhouse effect and essential validation of the water vapor greenhouse effect in climate models.

Earth radiation budget measurements are being used to study climate feedback mechanisms and to observe interannual variations and trends in the albedo and thermal emissions of Earth (Wong et al. 2006). Earth radiation budget measurements are now sufficiently well calibrated that long-term changes in the Earth's energy balance can be estimated from space-based measurements (Wielicki et al. 2002, 2005, Loeb et al. 2007). Long-term monitoring of Earth's energy balance allows greater understanding of the climate system's response to natural events such as El Niño and volcanic eruptions (see Box 4.3) and also may reveal aspects of the onset of human-induced global warming.

GLOBAL DISTRIBUTION OF CLOUD PROPERTIES

Knowledge of global distribution of cloud properties is required to understand the role of clouds in Earth's climate. Prior to the satellite era, observations of clouds were based on estimates made by human observers on the surface, providing only limited data coverage, particularly over the oceans. Beginning in the 1980s, an international climate research project under the World Climate Research Programme used satellite measurements taken for purposes of weather observation to create a data set of global cloud observations, giving the first estimates of the global distribution of cloud amount, optical depth, and cloud top temperature based on instrumental data (Schiffer and Rossow 1985, Rossow and Schiffer 1999). These results originate from the International Satellite Cloud Climatology Project, which continues today using a constellation of six operational geosynchronous (GEO) and low earth orbit (LEO) satellites. It is the longest continuous project using international satellites for climate monitoring.

Combining radiation budget measurements with cloud amount and type measurements from space has shown how different types of clouds contribute to the radiation budget, indicating that deep convective tropical clouds have a relatively small effect on the radiation balance of Earth but that marine stratocumulus clouds have a strongly negative impact on the radiation balance (Figure 4.5; Hartmann et al. 1992, Chen et al. 2000). The response of clouds to climate change remains one of the outstanding uncertainties in making projections into the future.

Estimates of global cloud properties from existing meteorological instruments are limited by the precision and spectral coverage of the instruments on the meteorological satellite platforms. New instruments with better calibration and more information about clouds are providing new opportunities to understand clouds and their role in climate. Moderate Resolution Imaging Spectroradiometer (MODIS) data provide much better calibration and spectral resolution than current or former meteorological satellites (King et al. 2003). Multiangle Imaging Spectroradimeter (MISR) data provide multiangle, multiwavelength visible views of clouds that can provide important information on cloud geometry and reflective properties (Diner et al. 2005). Measurements of clouds with cloud radar and light detection and ranging (lidar)

BOX 4.1
Total Solar Irradiance and Its Variability

The total solar irradiance, the total radiant energy coming from the Sun at the mean position of Earth, has been measured precisely from Earth-orbiting satellites for nearly 30 years, allowing the observation of nearly three solar cycles (Figure 4.2). To measure total solar irradiance precisely, it is important to remove the effect of the atmosphere's absorption, which can be achieved by taking the measurements from Earth orbit. Also, satellite orbits can be chosen to be in constant sunlight, allowing continuous monitoring of solar irradiance changes. These measurements show that the variation in total solar irradiance associated with the Sun's 11-year cycle is about 0.1 percent. Variations of 0.2 percent are associated with the Sun's 27-day rotation at times of high solar activity (Hickey et al. 1980, Willson et al. 1981, Willson and Hudson 1988, Frohlich and Lean 2004). These changes are small compared to the effect of greenhouse gases on the energy balance of Earth. It is important to monitor the energy exchange between Earth and space so that observed changes in Earth's climate can be attributed to and partitioned correctly among various causal mechanisms, including solar variability, atmospheric particles induced by volcanic eruptions, human-induced greenhouse gases, and aerosols.

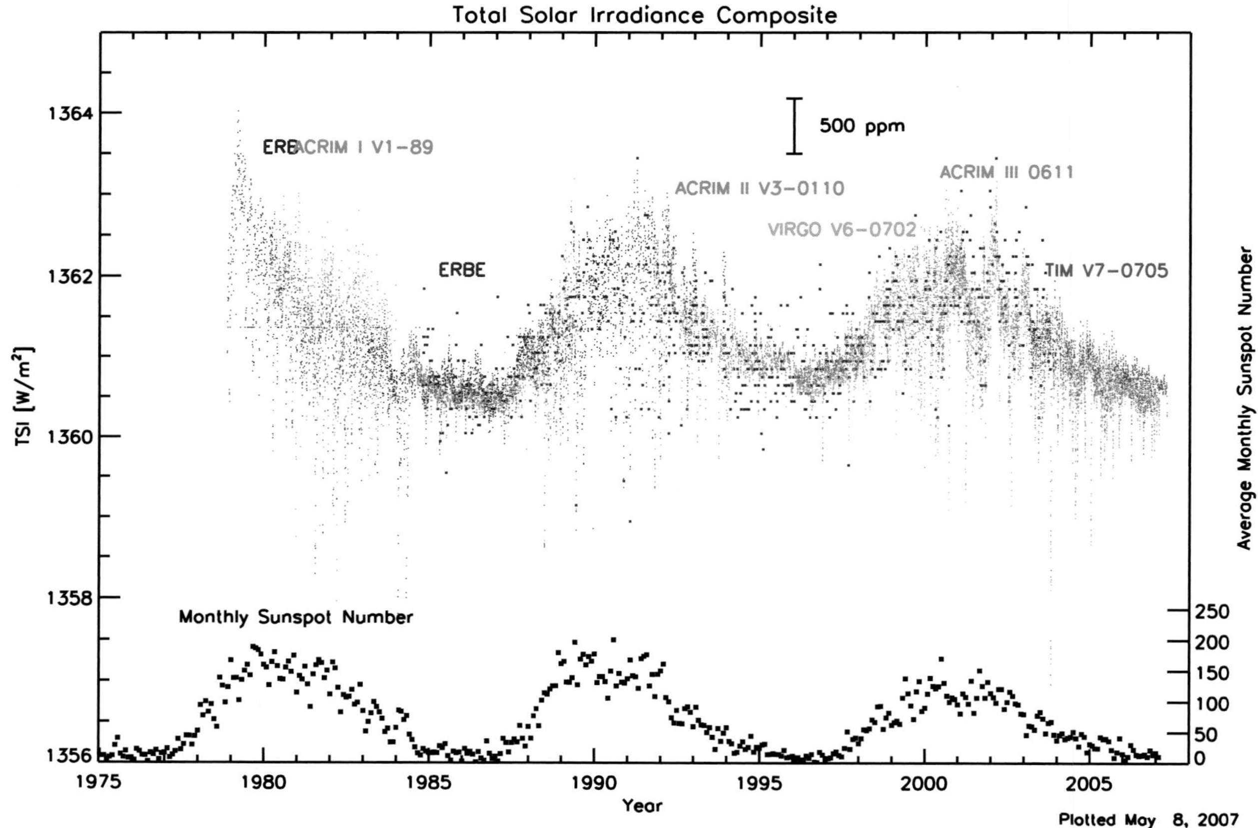

FIGURE 4.2 Time history of total solar irradiance (TSI) observed from seven different orbiting TSI monitors, along with monthly sunspot number. The average change in TSI during the solar cycle is about 1.5 W/m^2 or about 0.1 percent. SOURCE: Figure courtesy of Dr. Greg Kopp, University of Colorado, *http://spot.colorado.edu/~koppg/TSI/.*

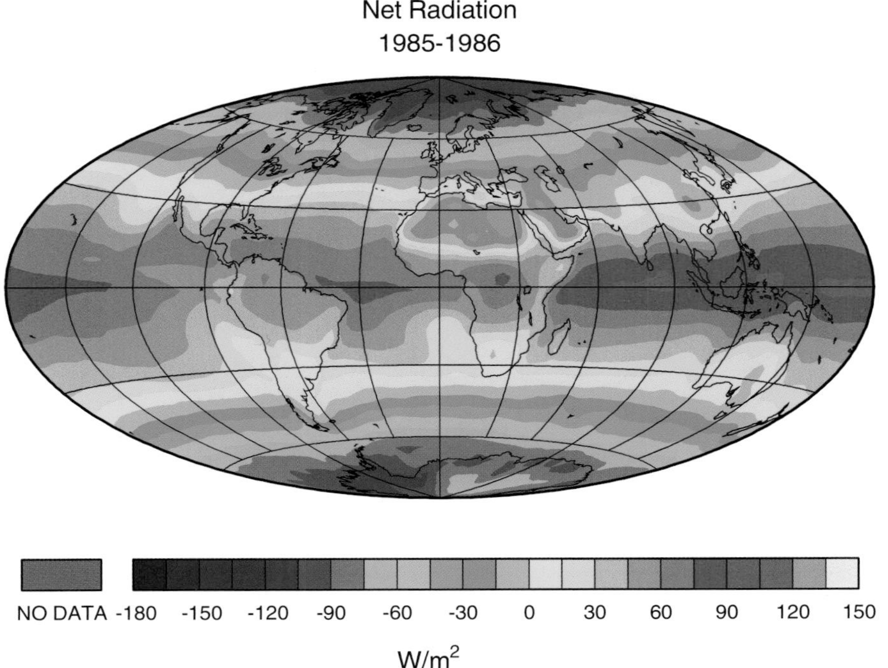

FIGURE 4.3 The annual mean net radiation balance from the Earth Radiation Budget Experiment (ERBE), 1985-1986. Positive values indicate net energy entering the Earth. In order to balance the energy budget, the atmosphere and ocean must transport heat from regions where the net input is positive to regions where it is negative. SOURCE: Graphic by D. Hartmann and M. Michelsen, University of Washington.

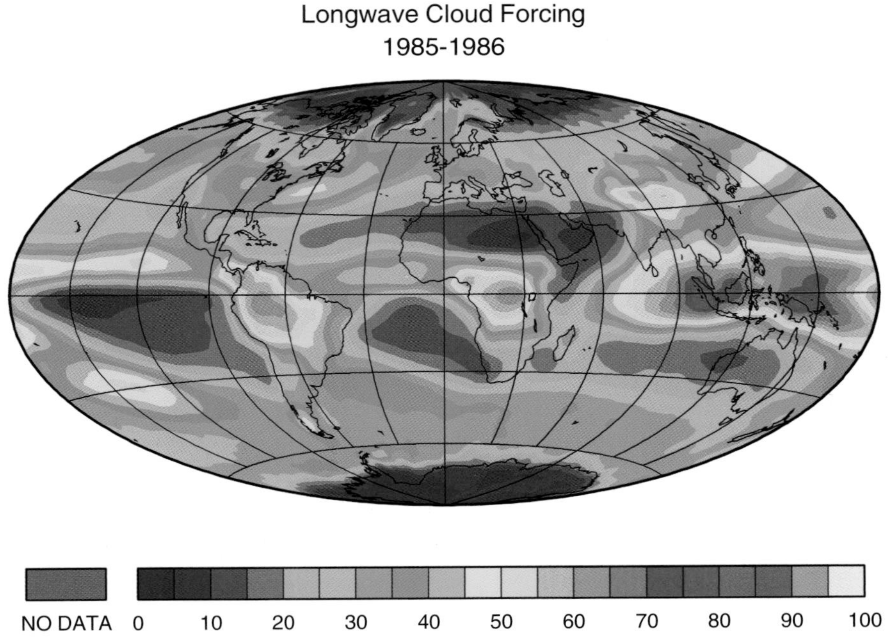

FIGURE 4.4 Longwave cloud forcing, the amount by which clouds reduce the escaping thermal emission from Earth, 1985-1986. Positive values indicate that clouds are reducing the thermal energy emission to space, a positive effect on the energy budget. Note the large positive forcing due to the deep convective clouds trapping longwave emission in the tropical West Pacific and Indian Ocean region and over the equatorial continents. SOURCE: Graphic by D. Hartmann and M. Michelsen, University of Washington.

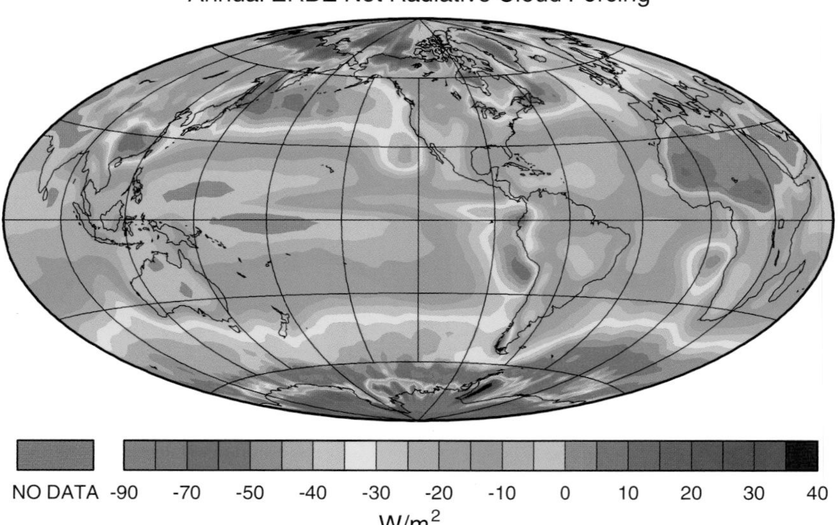

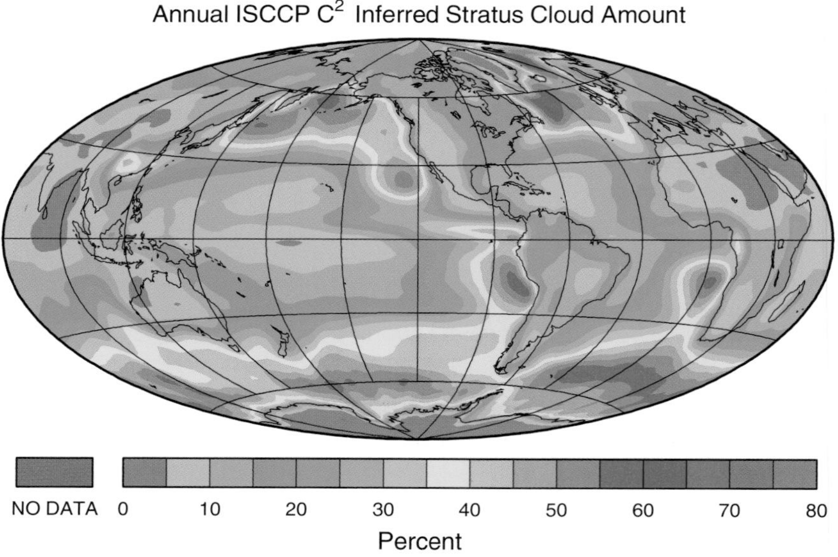

FIGURE 4.5 The top panel shows net cloud radiative forcing, annually averaged as observed by the ERBE. Negative values (red colors) indicate that clouds reduced the energy balance of Earth by reflecting more solar radiation than the amount by which they reduced the escaping infrared radiation. The bottom panel shows the fractional area coverage by low clouds as measured by the International Satellite Cloud Climatology Project (ISCCP). Note the close correspondence between low stratocumulus clouds over the ocean and strongly negative cloud radiative forcing. SOURCE: Graphic by D. Hartmann and M. Michelsen, University of Washington.

give unprecedented detail on vertical cloud structure. Cloud radar in space can provide good vertical resolution of reasonably thick clouds, including the tops and bottoms of layered clouds (Stephens et al. 2002). Lidar in space can provide very sensitive measurements of thin layers of clouds or aerosols (Winker 1997). These data have provided an unprecedented view of cloud structure, particularly in showing how clouds are layered vertically, which was not possible with visible, infrared, or microwave passive instruments.

The CloudSat radar and the Cloud Aerosol Lidar and Infrared Pathfinder Satellite Observations (CALIPSO) lidar provide a new dimension in observing the atmosphere. Rather than providing horizontal distributions of cloud and aerosol features typical of more conventional satellite sensors, these new nadir-pointing active sensors measure

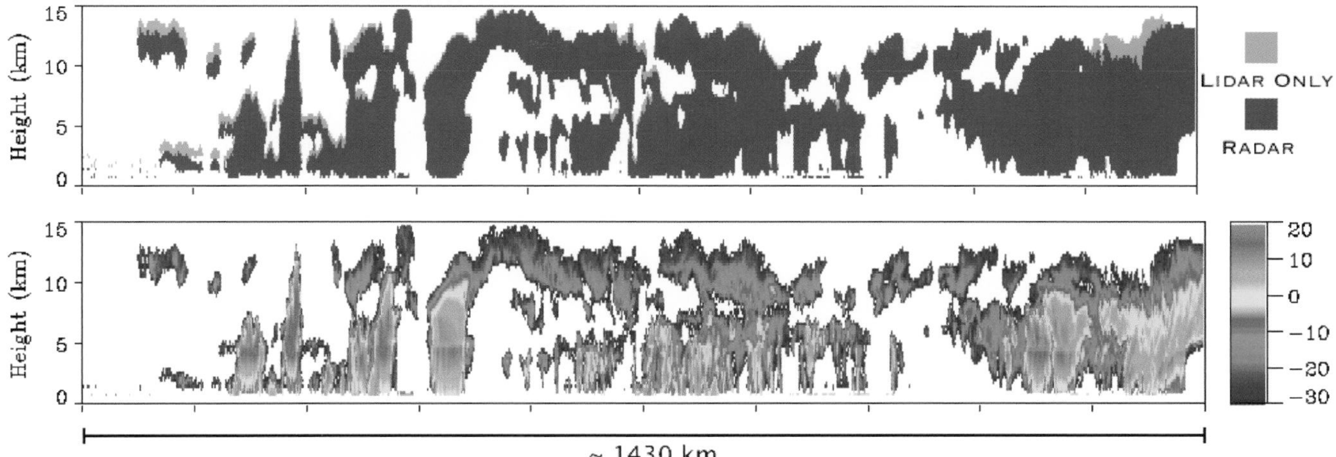

FIGURE 4.6 Portion of an orbit showing the cloud mask that combines CALIPSO lidar and CloudSat radar (upper panel) and the CloudSat radar reflectivity (lower panel). This example of monsoonal convection illustrates how precipitation (easily identified as regions of high reflectivity above the surface) falls from mixtures of deep and shallow convection. Shallow precipitating convection is often concealed from above by thick overlying cirrus clouds as apparent in the middle portion of this cross section. SOURCE: Image courtesy of G. Stephens.

the vertical structure of clouds and aerosols. The vertical structure revealed by CloudSat, for instance, offers deeper insights into the key processes that shape clouds and precipitation. For example, the image shown in Figure 4.6 is a cross section of the vertical distribution of radar reflectivity measured along a portion of one orbit. Also shown is the matching cloud mask information obtained from the combination of lidar and radar data. This example shows observations of clouds and precipitation associated with an active monsoon over southern China.

Observations such as these provide a way of observing the cloud structures with embedded precipitation and begin to provide hints about the way precipitation is organized. When accumulated over the entire tropics, these observations are now beginning to reveal that not all precipitation falls from deep convective clouds, as has generally been assumed, but that significant accumulations of water come from precipitation that falls from shallower clouds, as highlighted in this one example. This result has further implications for the nature of the vertical distribution of latent heating by precipitating cloud systems in the atmosphere, with ramifications on the way such clouds add (latent) heat to the atmosphere. The latter is essential for understanding the dynamic envelope of monsoons as well as the topic of the prediction of medium- and longer-term variability of the tropical atmosphere.

AEROSOLS FROM NATURAL PROCESSES AND HUMAN ACTIVITIES

An aerosol is a suspension of tiny liquid or solid particles in the atmosphere. Aerosol particles are distinguished from clouds by requiring that aerosol particles be stable in unsaturated air. Examples include dust, sulfuric acid particles, sea salt, organic particles, and smoke. Aerosols play important roles in the energy budget of Earth, in the formation of clouds, and in the chemistry of the atmosphere. Aerosol particles are produced naturally through biological emissions or elevation of particles by wind, but human activities provide a substantial enhancement to the natural aerosol loading of the atmosphere through agricultural and industrial activities. Aerosol particles can be produced either directly or by the chemical conversion of precursor chemicals that exist in solid or liquid form. Aerosols influence climate in several ways. Because aerosol particles reflect and absorb radiation, they can directly influence the energy balance of Earth. For many aerosols their primary effect is to reflect solar radiation and thereby cool the climate. Aerosols may also warm the atmosphere directly by absorption of radiation, however, and this is particularly important for highly absorbing aerosols such as soot (Figure 4.7).

Space measurements have succeeded in depicting aerosols associated with human activity over the oceans by isolating fine-mode from coarse-mode aerosols such as dust and sea salt that arise from natural processes. Plumes of fine aerosols are shown to result from biomass burning and from industrial activities (Tanré et al. 2001). The ability to distinguish fine from coarse aerosols has led to efforts to characterize the anthropogenic contribution to the aerosol direct forcing of climate (Bellouin et al. 2005, Kaufman et al. 2005).

INDIRECT EFFECTS OF AEROSOLS

Another way that aerosols can influence climate is through their role as the small particles on which clouds form (cloud condensation nuclei). An important contribution

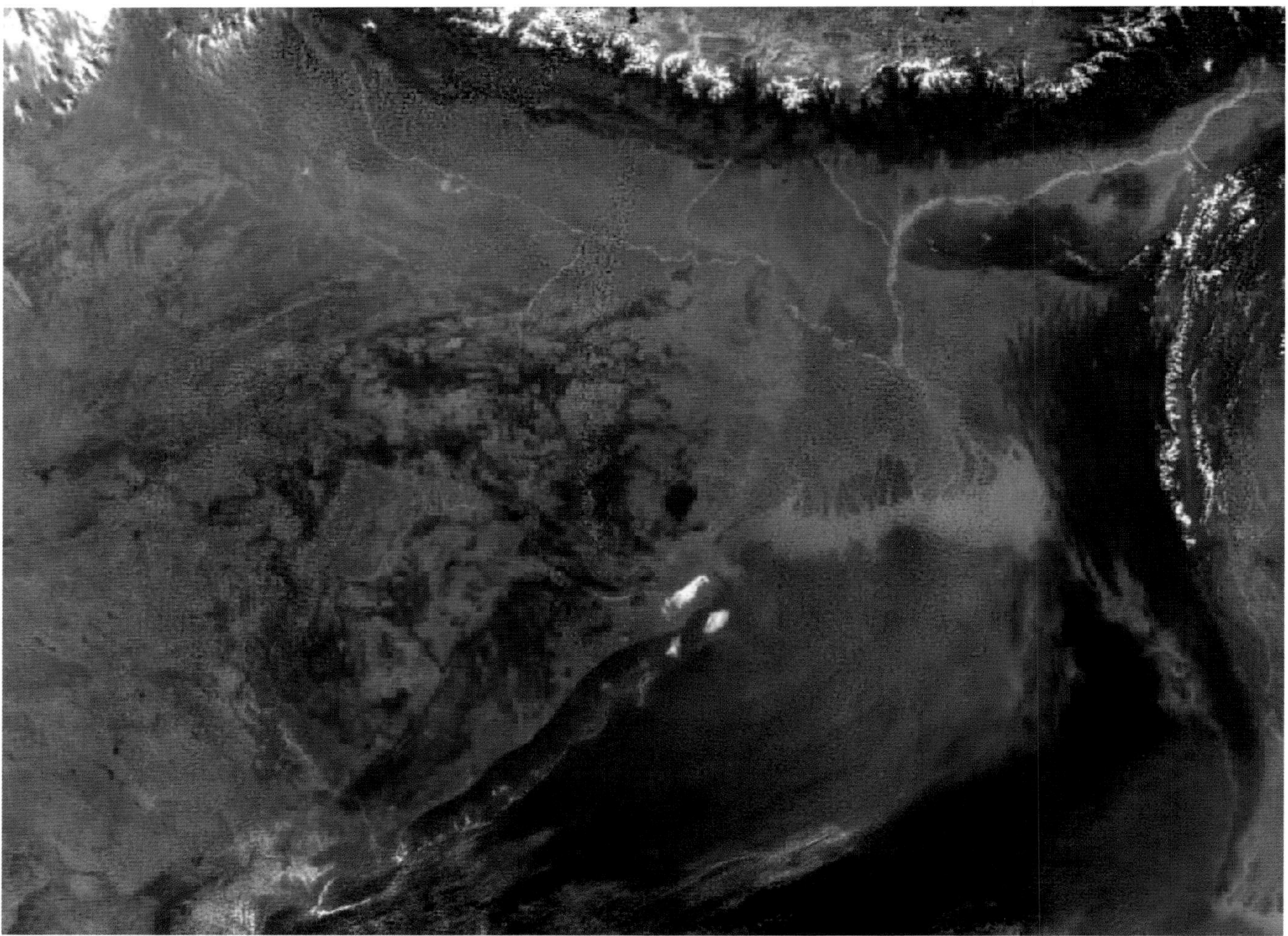

FIGURE 4.7 Against the arcing backdrop of the Himalayan Mountains (top of image), rivers of grayish haze follow the courses of the Ganges River and its tributaries (left) and the Brahmaputra River (right) on February 1, 2006. The plumes appear to combine like their watery counterparts and flow out together over the Bay of Bengal past the mouths of the Ganges, the multipronged delta of the river along the Bangladesh coast. This image was captured by MODIS on NASA's Terra satellite. Scientists studying the cloud of haze that frequently lingers over parts of Asia from Pakistan to China and even the Indian and Pacific oceans have called the pollution the "Brown Cloud." The mix of aerosols (tiny particles suspended in air) includes smoke from agricultural and home heating and cooking fires, vehicle exhaust, and industrial emissions. In addition to causing respiratory problems, the persistent haze appears to hinder crops by blocking sunlight and could be altering regional weather. SOURCE: NASA image created by Jesse Allen, Earth Observatory, using data obtained courtesy of the MODIS Rapid Response team, *http://visibleearth.nasa.gov/view_rec.php?id=20461.*

of satellite measurements to our understanding of the role of aerosols in climate was the discovery of the ship track phenomenon (Conover 1966, Coakley et al. 1987; Box 4.2, Figure 4.8). This heightened the awareness of the indirect effect human-produced aerosols have on the albedo of Earth.

STRATOSPHERIC PARTICLES

Stratospheric aerosols resulting from explosive volcanic eruptions and subsequent conversion of sulfur dioxide gas to aerosols have been observed by satellite for the El Chichon and Pinatubo eruptions (McCormick 1992, Lambert et al. 1993, McCormick et al. 1993; Box 4.3, Figure 4.9). The initial development, transport, mixing, and gradual decline of the aerosols associated with these eruptions provide a sound basis for understanding the effect of such volcanic eruptions on surface climate and better estimates of the likely effects of large volcanic eruptions that have occurred in the more distant past. Reflected solar measurements allow the evolv-

BOX 4.2
Discovery of the Ship Track Phenomenon

Ship tracks can be observed in the atmosphere because very small airborne particles emitted in the exhaust of large ships attract water molecules, acting as cloud condensation nuclei, and leave bright streaks in the air after the ships have passed. Ship tracks are a visible example of how human-produced aerosols can indirectly change the energy balance of Earth by changing the properties of clouds by acting as cloud condensation nuclei (Figure 4.8). This indirect effect of clouds is currently one of the major uncertainties in computing the effect of human activities on Earth's climate.

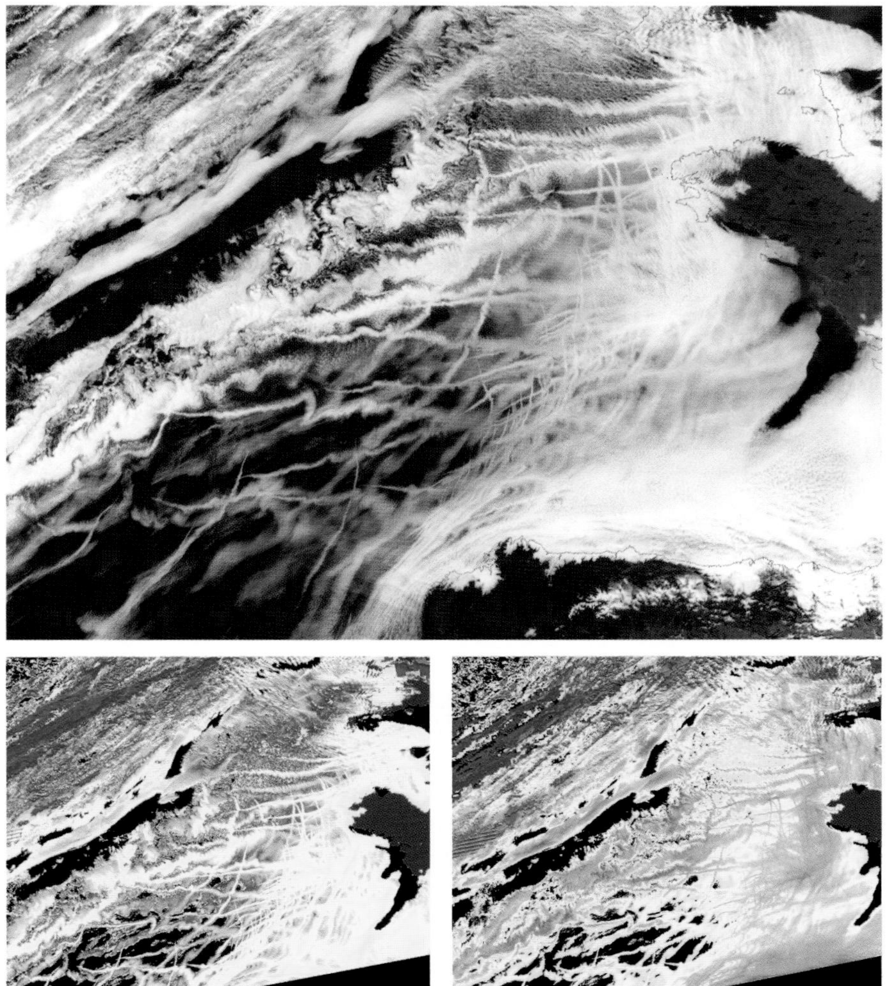

FIGURE 4.8 The top panel shows a true color image from the MODIS instrument taken over the Atlantic Ocean on January 27, 2003. Bright linear features are apparent in the low clouds in much of the scene. MODIS can independently measure the optical depth (lower left panel), which is enhanced in the bright regions, and the effective particle radius (lower right panel). The smaller particle radius in the ship tracks is what would be expected from the introduction of many more cloud condensation nuclei from the ship exhaust. Smaller particles are more effective in reflecting solar radiation. This strongly suggests that the cloud enhancements are caused by human-produced aerosols. SOURCE: Images courtesy of Jacques Descloitres, MODIS Land Rapid Response Team, and Mark Gray, MODIS Atmosphere Science Team, both at NASA Goddard Space Flight Center, *http://earthobservatory.nasa.gov/Newsroom/NewImages/images.php3?img_id=11271*.

BOX 4.3
Response of Earth's Radiation Budget to a Volcanic Eruption

The response of the radiation balance to the eruption of Mount Pinatubo was directly measured with broadband radiation instruments on Earth-orbiting satellites. This allowed not only a direct confirmation of the effect of stratospheric volcanic aerosols in reducing the energy balance but also verification of the model-predicted surface cooling in response to the eruption, giving additional confidence in our ability to model climate variability and change (Hansen et al. 1992, Minnis et al. 1993, Soden et al. 2002; Figure 4.9). The stratospheric aerosols resulting from the eruption were independently measured from Advanced Very High Resolution Radiometer (AVHRR) data and Stratospheric Aerosol and Gas Experiment SAGE data (McCormick et al. 1995). This is a prime example of how the length and continuity of a given data record yields additional scientific benefits beyond the initial research results of the mission and beyond the monitoring implications for operaitonal agencies.

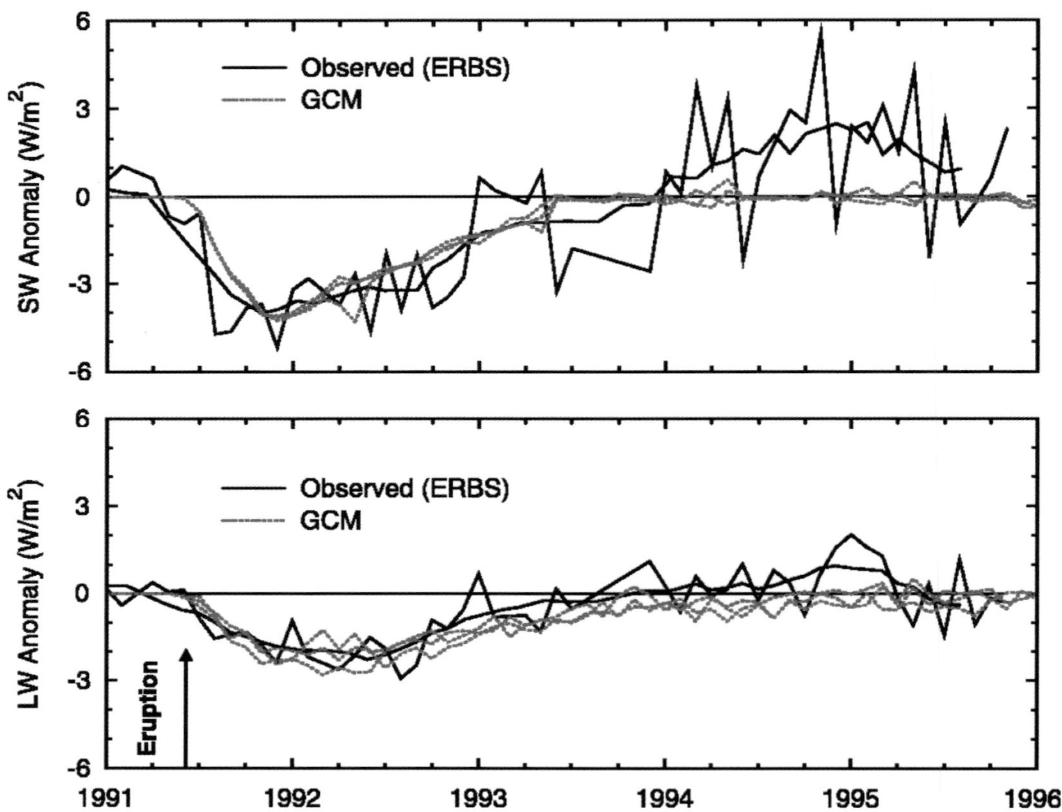

FIGURE 4.9 Comparison of the observed anomalies in absorbed shortwave (top panel) and emitted longwave (bottom panel) radiative fluxes at the top of the atmosphere from ERBE satellite observations (black) and three ensembles of Global Climate Model (GCM) simulations (red). The observed anomalies are expressed relative to a 1984-1990 base climatology, and the linear trend is removed. The GCM anomalies are computed as the difference between the control and Mount Pinatubo simulations for each ensemble member (the Mount Pinatubo eruption of May 1991 is marked on the bottom panel). The results are expressed relative to the preeruption (January to May 1991) value of the anomaly and smoothed with a 7-month running mean (thick line). Both the model and the observed global averages are from 60° N to 60° S due to the restriction of observed data to these latitudes. SOURCE: Soden et al. (2002). Reprinted with permission from AAAS, copyright 2002.

ing radiative impact of the aerosol cloud to be measured (Stowe et al. 1992).

Mineral aerosols are important for the trace metal balances of the ocean, which in turn are important for ocean biology (see Chapters 8 and 9). Satellite images show the dramatic export of mineral aerosols from the Sahara Desert to the Atlantic Ocean during dust storms (Figure 4.10). Mineral dust can also have a significant impact on climate.

GLOBAL CLIMATOLOGIES OF AEROSOLS

Aerosol concentrations vary strongly over time and space, and quantifying the various effects of aerosols requires continuous global measurement, which can best be achieved from Earth-orbiting satellites (Figure 4.11). A major contribution of Earth observations from space is the development of global climatologies of aerosols. These have been obtained from visible measurements from weather satellites (Stowe et al. 1997, Nakajima and Higurashi 1998, Mishchenko et al. 1999), from ultraviolet measurements from the Total Ozone Mapping Spectrometer (Herman et al. 1997, Torres et al. 2002), and from the instruments on the Earth Observing System suite of instruments, especially MODIS (Chu et al. 2002, Remer et al. 2002, 2005) and MISR (Kahn et al. 2005; Figure 4.11). Aerosol properties can also be inferred from polarization (Tanré et al. 2001) and from lidar measurements from space (Winker et al. 1996). Global aerosol measurements from space are greatly improved by their validation with surface sun photometer measurements (Holben et al. 1998, Dubovik et al. 2000).

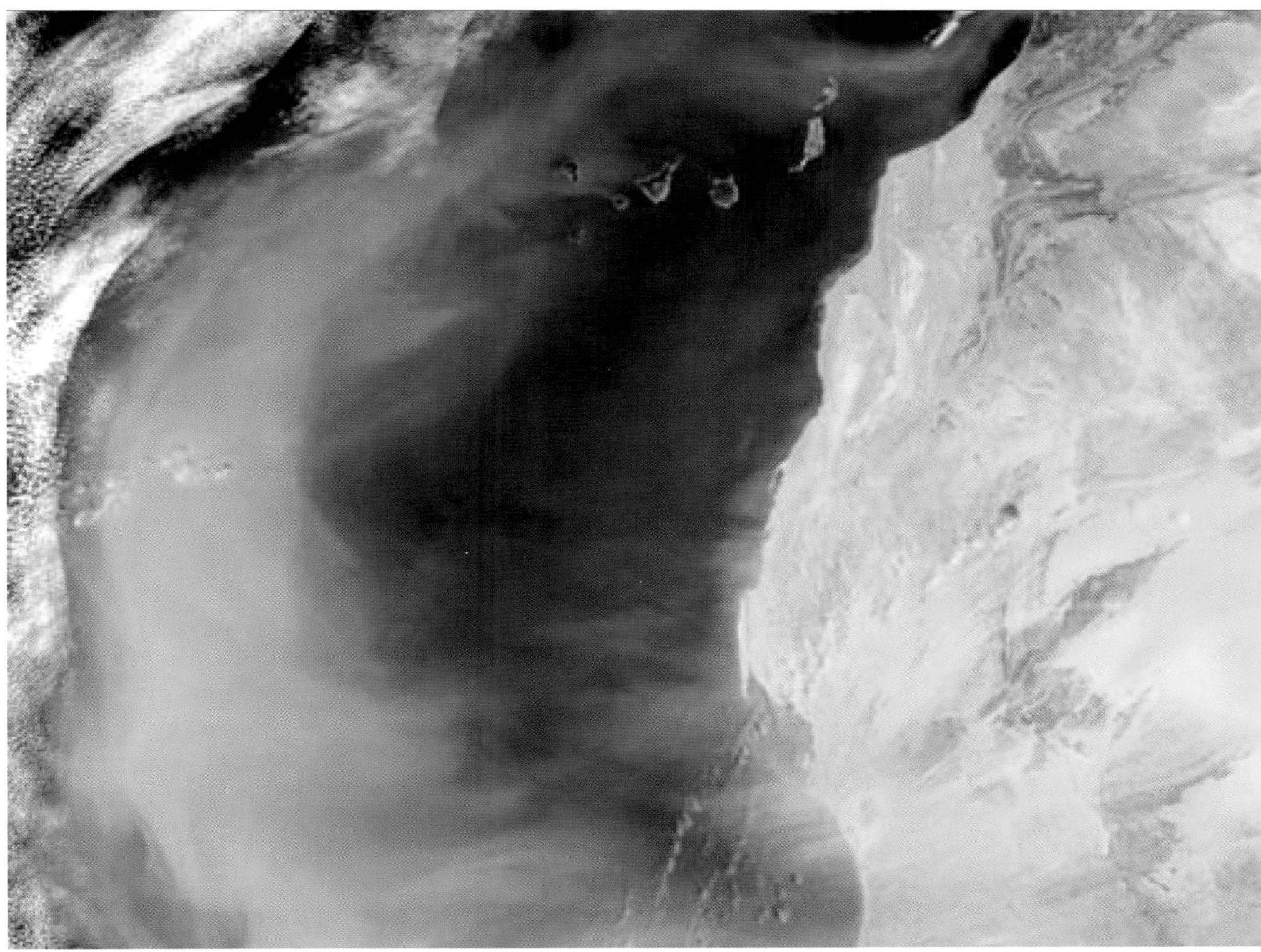

FIGURE 4.10 An intense African dust storm sent a massive dust plume northwestward over the Atlantic Ocean on March 2, 2003. In this true-color scene, acquired by MODIS aboard NASA's Terra satellite, the thick dust plume (light brown) can be seen blowing westward and then routed northward by strong southerly winds. The plume extends more than 1,000 miles (1,600 km), covering a vast swath of ocean extending from the Cape Verde Islands (lower left), off the coast of Senegal, to the Canary Islands (top center), off the coast of Morocco. SOURCE: Image courtesy of Jacques Descloitres, MODIS Rapid Response Team, NASA GSFC, *http://visibleearth.nasa.gov/view_rec.php?id=5619.*

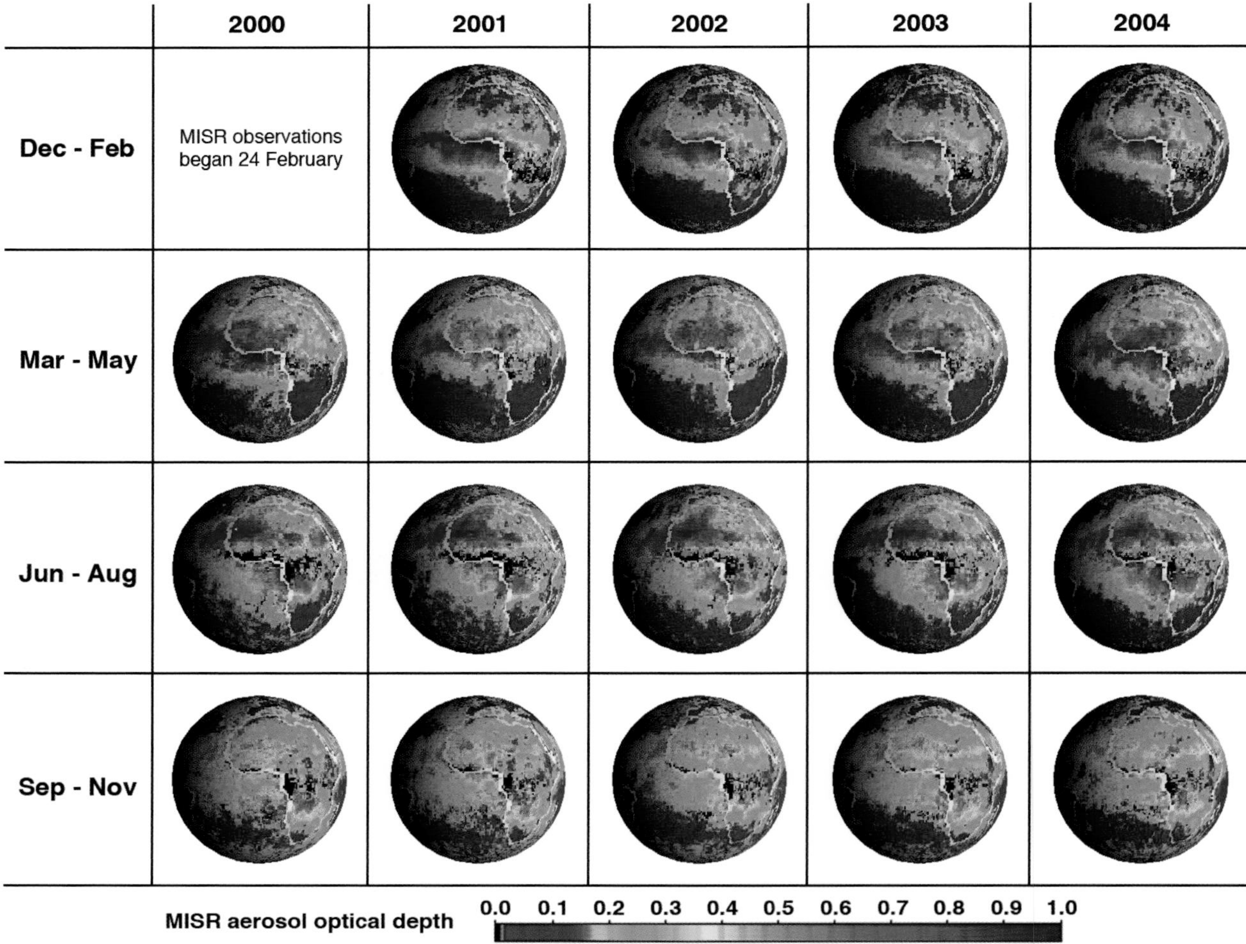

FIGURE 4.11 These 19 global panels show the seasonal average distribution of atmospheric aerosol amounts across Africa and the Atlantic Ocean. The measurements capture airborne particles in the entire atmospheric column, for subvisible sizes ranging from tiny smoke particles to "medium" dust (about 0.5 to 2.5 microns). Such particles are produced by forest fires, deserts, volcanoes, breaking ocean waves, and urban and industrial pollution sources. SOURCE: NASA, GSFC, Langley Research Center (LaRC), Jet Propulsion Laboratory (JPL), Multiangle Imaging Spectroradimeter (MISR) Team.

Space observations have the potential to allow the estimation of the global average optical depth of aerosols, which is presently unknown. Rapid global coverage also allows sources of aerosols to be inferred from plumes of aerosols that can be observed over the oceans (Herman et al. 1997, Husar et al. 1997). Multiyear records of aerosol optical depth over water show reproducible seasonal patterns (Torres et al. 2002). Measurements from space show a surprisingly large contribution from Saharan dust and biomass burning and distinct differences between the northern and southern hemispheres, presumably due to human production of aerosols (Husar et al. 1997, Prospero et al. 2002).

The ability to distinguish aerosols from clouds and fine aerosols from coarse aerosols combined with the ability to construct a long-term record of aerosols is a remarkable accomplishment and demonstrates how sophisticated satellite technology and analysis tools have become. This newly gained observational capability greatly enhances our understanding of climate forcing by aerosols from natural and anthropogenic sources and leads to improvements in climate modeling.

5

Atmospheric Composition: Ozone Depletion and Global Pollution

Although on average there are only four molecules of ozone for every 10 million molecules of air, it is central to atmospheric composition for several reasons. Stratospheric ozone absorbs ultraviolet (UV) radiation; thus, it shields the lower atmosphere and Earth's surface from UV radiation that is harmful to living organisms. This absorption of UV radiation warms the stratosphere and plays a major role in establishing the temperature structure of the atmosphere (Box 5.1), while its infrared (IR) absorption and emission are also important in Earth's energy balance. No less important are the chemical and photochemical reactions of ozone with other species, which regulate the trace gas structure of the stratosphere and troposphere. Given ozone's importance to the atmospheric composition and temperature structure and as Earth's UV shield, it is critical to measure its global distribution; understand its trends and the mechanisms that control its distribution; and understand its interactions with atmospheric chemistry, dynamics, and the climate system.

A delicate balance of photochemical reactions among oxygen (molecular and atomic), nitrogen oxides, hydrogen oxides, and halogenated oxides maintains the "ozone layer" in the stratosphere, where 90 percent of the total ozone column resides. A rapid increase in anthropogenic halogen-containing gases (collectively called "halocarbons," which includes chlorofluorocarbons [CFCs]) over the past 50 years resulted in a huge perturbation to the natural stratospheric ozone balance.

Satellite-generated data have made a vital contribution to understanding the threat of stratospheric ozone depletion, observing stratospheric dynamics, and determining the cause for the Antarctic ozone hole. Satellite observations provided the first global measurements of stratospheric ozone and temperatures, expanding and revolutionizing our understanding of the atmosphere above the tropopause. This knowledge confirmed the dangers associated with the release of anthropogenic CFCs and other halocarbons and helped shape international policies to minimize their use and release.

In contrast to the stratosphere, the elevated ozone levels in the troposphere are problematic. The gas-phase composition of the troposphere is very complex, bound by the tropopause above and the ocean and land surface below (Figure 5.1). In tropical and midlatitudes especially, the ocean and land surfaces are active sources of trace gases that are linked through physical and chemical transformations to other tropospheric species. Tropospheric ozone together with other radicals (most importantly hydroxide) contributes to the oxidative capacity. This oxidative capacity is important to the removal of the most reactive air pollutants, which cleanses the atmosphere. Similar to its stratospheric counterpart, tropospheric ozone levels are strongly influenced by photochemical reactions involving nitrogen oxides and hydrogen oxides and by ozone inputs from the stratosphere. In the presence of elevated nitrogen oxides from localized sources of pollution, hydrocarbons—even the ubiquitous methane and carbon monoxide—can perturb the natural ozone balance toward unhealthy levels. Ozone's powerful oxidizing capacity threatens human health, agricultural productivity, and natural ecosystems. In addition, ozone in the free troposphere is a powerful greenhouse gas. In contrast to the stratosphere, halogen chemistry is relatively benign in the troposphere, with a noteworthy exception in polar regions.

Stratospheric ozone studies have benefited greatly from a long history of space observations. These investigations represent one of the best-known examples of satellite "successes" in Earth observations. The role that satellites have played in tropospheric ozone is more complex but no less important. As this chapter documents, satellites discovered ozone "pollution" in the remote tropics 20 years ago. Follow-on exploration with newer satellites and ground-based and aircraft instrumentation has shown that climate dynamics in the tropics and stratospheric forcing can be as significant as photochemical reactions. Today, a new generation of satellite instrumentation, described later in this chapter, is mapping tropospheric ozone globally along with its photochemical relatives (nitrogen dioxide bromide, carbon monoxide, formaldehyde).

BOX 5.1
Atmospheric Structure

The atmosphere is often divided into layers to differentiate regions with different characteristics. The lowest layer is called the troposphere, from Greek words indicating a region of overturning. In this region the temperature generally decreases with altitude to a height called the tropopause. Above this altitude lies the stratosphere, where the temperature remains constant or increases with altitude up to the stratopause, at about 50 km altitude. The temperature again decreases in the overlying mesosphere up to the mesopause (~80 km), above which it rises in the thermosphere under the influence of solar radiation. The temperature structure in the stratosphere tends to suppress vertical motions, leading to more horizontal winds.

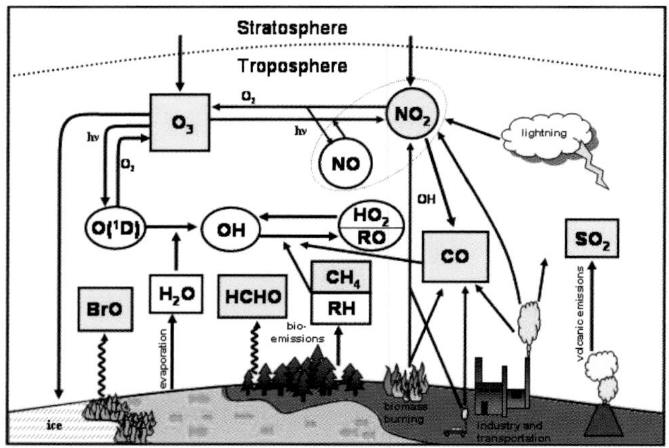

FIGURE 5.1 Schematic of biogeochemical cycling with human contributions included, illustrating the major gas-phase constituents in the lower atmosphere that are measured from space (shaded in gray). NO_2 and NO are encircled to represent equilibrium; their sum is referred to as NO_x. NOTE: BrO = bromine monoxide; CH_4 = methane; CO = carbon monoxide; h = Planck's constant; H_2O = water; HCHO = formaldehyde; HO_2 = hydroperoxyl; ν = photon frequency; NO = nitric oxide; NO_2 = nitrogen dioxide; $O(^1D)$ = electronically excited oxygen atoms; O_2 = molecular oxygen; O_3 = ozone; OH = hydroxide; RH = hydrocarbon species; SO_2 = sulfur dioxide. SOURCE: Drawing by A.M. Thompson and K.M. Dougherty, Pennsylvania State University.

UNDERSTANDING AND REMOVING THE THREAT OF STRATOSPHERIC OZONE DEPLETION

Until about 1964 it was thought that the Chapman (1930) scheme, based only on forms of oxygen, could explain the stratospheric abundance of ozone (Wayne 1985). Subsequently, improved laboratory measurements of reaction rate coefficients showed that this approach overestimated the amount of stratospheric ozone by a factor of 2. Further laboratory measurements showed that reactions involving compounds of hydrogen, chlorine, and bromine from natural sources could enter into catalytic cycles that would speed up the rate of ozone destruction, decreasing estimates of ozone amounts and bringing them into better agreement with ground-based observations.

A very alarming fact was that two of these gases had large and potentially rapidly increasing anthropogenic sources. A projected fleet of 500 commercial supersonic airplanes flying many hours each day was expected to inject large amounts of nitrogen oxides into the lower stratosphere, with deleterious effects (Crutzen 1970, Johnston 1971). Ultimately this fleet did not materialize. However, the techniques and models developed to address the former problem were ready to be applied to the next threat to the ozone layer: chlorine, which was being released in significant amounts by the photolysis of CFCs in the stratosphere. The chlorine released from CFCs was also predicted to cause a serious reduction in ozone (Cicerone et al. 1974, Molina and Rowland 1974). These gases would reduce the amounts of stratospheric ozone below the natural background level, letting more UV radiation reach the surface, causing increased incidence of human skin cancer as well as damage to other biological processes. Subsequent studies showed that bromine, which has natural and anthropogenic sources, could also cause significant ozone depletion (Wofsy et al. 1975, Yung et al. 1980). Because of the dire nature of these predictions, it was crucial to develop a better understanding of this region of the atmosphere as quickly as possible.

OBSERVING STRATOSPHERIC DYNAMICS

To predict the ozone distribution and its changes in the stratosphere, it is also necessary to understand atmospheric motions. These are closely linked to radiative heating and cooling, which depends on the atmospheric composition, notably the ozone distribution (Craig 1965). Understanding this interacting system of chemistry dynamics, and radiation requires global observations, unavailable from ground-based measurements, as well as the synergistic use of models to incorporate this information and allow accurate and trustworthy predictions to be made.

The stratosphere was first identified in 1899, when balloonborne measurements showed that the atmospheric temperature did not continue to decrease with altitude but became constant or even increased above a height termed the tropopause. By the end of World War II, data from operational weather balloons, which could in the best case reach altitudes of 30 km, provided a picture of stratospheric temperatures and winds up to this altitude but with low spatial and temporal resolution. After World War II, rocket soundings at about a dozen locations extended these measurements to 65 km and occasionally higher. These data were enough to delineate the global variation of the vertical temperature and wind distribution but yielded little information on features with smaller temporal or horizontal scales and provided little information on the southern hemisphere's atmosphere. Only the roughest picture of motions at the longest horizontal scales was available, and many studies were constrained by data availability to limited altitude and geographic regions. Craig (1965) presents a good discussion of knowledge and speculation at that time.

Similarly, a limited number of ground-based instruments in the pre-satellite era, mainly UV spectrophotometers, could provide only a rudimentary view of the vertical, latitudinal, and seasonal variations of the ozone distribution, with large uncertainties (Goody 1954). One particular puzzle was the nature of the atmospheric motions that transport ozone from high altitudes in the tropics, where it is produced by solar UV radiation and atmospheric chemistry, to low altitudes in polar regions and midlatitudes, where processes destroying it dominate (Craig 1965).

It was recognized that the transport of ozone and other gases, as well as heat and momentum, was important. Initially, such questions were addressed in terms of conventional fluid dynamics (Hunt and Manabe 1968). However, these early estimates were so uncertain that it was often impossible even to determine whether the ozone transport was northward or southward (NRC 1979).

The understanding of atmospheric dynamics was revolutionized beginning in 1969, when satellite instruments were launched to measure temperature and ozone in the stratosphere. The first downward- or nadir-looking temperature sounders on Nimbus 3 (Box 5.2) demonstrated that remote sounding techniques could provide global observations of atmospheric temperatures from the surface to mesospheric altitudes. Although the soundings from these and subsequent nadir-looking instruments had low vertical resolution, they were sufficient to allow the heights of atmospheric pressure surfaces to be calculated. The balance between the slopes of these heights and Earth's rotation allowed scientists to make accurate calculations of stratospheric winds (Smith and Bailey 1985). These winds could be separated into winds in the east-west direction (along parallels of latitude) and into north-south wavelike perturbations. A review of early results clearly showed that the stratosphere and mesosphere were dynamically very active, with large (planetary)-scale waves propagating from the troposphere into the stratosphere and mesosphere in the winter hemisphere (Hirota and Barnett 1977).

The advent of horizon-viewing sounders provided temperature measurements with higher vertical resolution. These more detailed pictures revealed a wider range of atmospheric motions, including waves in the tropics with short vertical wavelengths, and provided more detail on planetary wave activities. The accurate and densely spaced measurements of temperature and the derived estimates of wind speed made it possible to study global transport in more detail. To avoid the uncertainties associated with conventional fluid dynamics, theoreticians recast their equations to essentially follow air parcels, leading ultimately to simple but accurate approximations. These showed how planetary waves, propagating up from the troposphere, could interact with the eastward wind motions, and thereby change the mean vertical and poleward circulation (Matsuno 1971, Andrews and McIntyre 1976). This explained the phenomenon of "sudden stratospheric warming," in which temperatures in the polar stratosphere at an altitude of 30 km can increase by 30°C or more in a few days.

One particular scientific achievement should be noted. Brewer (1949) and Dobson (1956) had independently postulated a mean north-south overturning circulation in the stratosphere, in which air rises from the troposphere into the stratosphere in the tropics and then travels to high latitudes (in both hemispheres) where it returns to the troposphere. Observations of distributions of methane, nitrous oxide, ozone, and water vapor (all from the limb sounders) were used to test and validate these (then-novel) ideas and theoretical approaches to the calculation of net transport of these gases (Andrews et al. 1987). A related triumph was the observation of the tropical "tape recorder" (Mote et al. 1996), which significantly advanced and confirmed scientific understanding of stratospheric dynamics and motions (Box 5.3).

Remote sounding also provided information on the composition of the stratosphere. The first instruments to shed light on the global distribution of stratospheric ozone were the backscattered ultraviolet (BUV) on Nimbus 4 and the solar backscattered ultraviolet (SBUV) and the Total Ozone Mapping Spectrometer (TOMS) on Nimbus 7, which provided global measurements of the total ozone in a vertical column. In addition, the Limb Radiance Inversion Radiometer (LRIR), the Limb Infrared Monitor of the Stratosphere (LIMS), and the Stratospheric and Mesospheric Sounder (SAMS) on Nimbus 6 and 7 retrieved the distributions of ozone, water vapor, nitrogen dioxide, nitric acid, nitrous oxide, and methane. The power of these measurements is shown by the observations of nitrogen dioxide and nitric acid, present in concentrations on the order of only 10 parts per billion by volume.

DETERMINING THE CAUSES OF ANTARCTIC OZONE DEPLETION

The advances in our understanding of the cause and dynamics of the Antarctic ozone hole exemplify the productive interactions of satellite observations with in situ and ground-based observations and numerical models. As

BOX 5.2
Remote Sensing of the Stratosphere[a]

Earth and its atmosphere emit infrared (or heat) and microwave radiation to space, which can only be detected from Earth-orbiting sensors. The emerging signal at wavelengths corresponding to the absorbing bands of atmospheric gases will depend on the vertical distribution of the gas, the strength of its absorption, and the temperature at a level where the chance of a photon escaping to space is about 37 percent. Because the fraction of carbon dioxide with altitude is known and nearly constant in the atmosphere, Louis Kaplan (1959) suggested that the vertical temperature profile could be retrieved by measuring the radiation emerging from the atmosphere as a function of wavelength in the 15-μm bands of carbon dioxide. Satellite Infrared Spectrometer (SIRS-A, a filter radiometer) and the Infrared Interferometer Spectrometer (IRIS-A) on Nimbus 3 (launched in 1969) were the first two instruments to demonstrate that remote sounding techniques could provide global observations of atmospheric temperatures from the surface into the stratosphere. Improvements of these instruments were accompanied on Nimbus 4 (1970) by the Selective Chopper Radiometer (SCR), which filtered the radiance through cells of carbon dioxide and permitted temperatures into the mesosphere to be determined.

Soundings from these and subsequent downward-looking (sometimes referred to as nadir-viewing) instruments had low vertical resolution. By looking at the horizon or limb with a narrow field of view, much higher vertical resolution (2-4 km) can be achieved (Gille and House 1971). The first instrument for infrared measurements was the Limb Radiance Inversion Radiometer (LRIR) on Nimbus 6 (1975). The Limb Infrared Monitor of the Stratosphere (LIMS) and the Stratospheric and Mesospheric Sounder (SAMS) followed on Nimbus 7 (1978). An additional advantage of this technique is that it allows measurement of trace gases, from cloud tops into the mesosphere, also with high vertical resolution. In 1991 the Microwave Limb Sounder (MLS) on the Upper Atmosphere Research Satellite (UARS) exploited the limb sounding geometry to measure additional species that emit measurable signals in the millimeter-wavelength part of the microwave spectrum.

Some of the sunlight striking the atmosphere is scattered back toward space and absorbed by atmospheric gases in its path. The spectral distribution of this backscattered radiation is affected by the amounts and vertical distributions of the absorbing gases. The nadir-viewing backscattered ultraviolet (BUV) instrument on Nimbus 4 was the first orbiting instrument to make use of these principles to determine the ozone distribution as a function of altitude and the total amount in a vertical column. The solar backscattered ultraviolet (SBUV) and the Total Ozone Mapping Spectrometer (TOMS) instruments were first flown on Nimbus 7. TOMS, by scanning from side to side, provided maps with complete global coverage of the total column amounts of ozone. Although designed for a year's operation, it lasted from 1978 until 1993, providing an excellent and consistent long-term data record of ozone columns.

Occultation measurements are another method for sounding the atmosphere. In this case the instrument measures light from the Sun as it sets behind the atmosphere as seen from the satellite. Trace gases absorb the radiation as it passes through the atmosphere, and aerosols scatter sunlight from the beam. The vertical distributions of these gases and aerosols can be determined by measuring the decrease of sunlight as a function of altitude in spectral bands where atmospheric gases absorb. The second such experiment was the Stratospheric Aerosol and Gas Experiment. The Halogen Occultation Experiment (HALOE) on UARS was designed to measure trace gases that were otherwise difficult to observe.

[a]Described by Houghton et al. (1984).

satellite observations were clarifying the structure, dynamics, and composition of the stratosphere, modeling activities were advancing rapidly. After about a decade of evolution, these models were predicting relatively modest but still important decreases in ozone concentrations, centered near 40 km altitude (Wuebbles et al. 1983). Later data confirmed these expectations (Solomon 1999).

However, in 1984 the world was startled by the discovery of a much larger than predicted ozone decrease over Antarctica at a much lower altitude, near 20 km in the lower stratosphere (Farman et al. 1985). This feature quickly became known as the Antarctic ozone hole. This unexpected phenomenon called into question the theoretical understand-

BOX 5.3
Discovery of the Tropical Tape Recorder

The discovery of the so-called tape recorder (Mote et al. 1996) represents a remarkable scientific achievement in understanding stratospheric dynamics and motions. The temperature of the tropical tropopause controls the fraction of water vapor in the air at the tropopause, near 16 km. Colder temperatures during a northern hemisphere winter "freeze dry" the air to a greater extent than the warmer temperatures later in the year. Figure 5.2, based on data from the Microwave Limb Sounder (MLS) on Aura, shows periods of dry air (negative departures from the mean) alternating with periods of more moist air (positive departures) at each level. These create alternating bands sloping upward, confirming the rising motion in the tropics. A closer look indicates that the slope of the bands, proportional to the upward velocity, varies with season. This varies with the convergence of wave activity in the upper stratosphere, as expected from theory. The tropopause thus acts like a recording head, with the temperatures "recording" the time-varying water vapor amounts on the air.

FIGURE 5.2 Time series of zonal mean water vapor profile measurements by the Microwave Limb Sounder on the Aura satellite. The colors represent a percentage change relative to the 15° S-15° N mean at each pressure level. The upward progression with time above the 140-hPa level (~14 km altitude) shows the vertical motion consistent with theoretical predictions. SOURCE: Figure courtesy of Jonathan Jiang, National Aeronautics and Space Administration, Jet Propulsion Laboratory.

ing of the mechanisms of ozone destruction and the model projections.

Satellite measurements played two important roles in unraveling the questions of ozone depletion:

1. measurements of trace species that lead to or catalyze ozone destruction contributed to confirming the causes of the depletion
2. measurements of stratospheric ozone concentrations and distribution, and their changes over time, enabled comparison to model predictions.

Theory predicted that the tropical upwelling—discussed in Box 5.2—carried CFCs and other halocarbons from the tropospheric source into the stratosphere, where they were present at only a few molecules per 10 billion atmospheric molecules. In the stratosphere, solar UV breaks CFCs apart, releasing chlorine molecules that react to produce relatively inert hydrochloric acid (HCl). Laboratory investigations showed that HCl could react on the surface of polar stratospheric clouds (discovered by the Stratospheric Aerosol and Gas Experiment [SAGE] measurements; McCormick et al. 1982) releasing approximately 1 part per billion of chlorine monoxide (ClO). ClO effectively catalyzes ozone destruction in the presence of sunlight.

Images from the second generation of infrared limb sounders confirmed the transport of these extremely small amounts of CFCs into the stratosphere and also the presence of the predicted compound chlorine nitrate ($ClONO_2$; Nightengale et al. 1996, Mergenthaler et al. 1996). In addition, Halogen Occultation Experiment (HALOE) measurements demonstrated the amounts and distribution of HCl (Russell et al. 1996). The picture was complete when sensitive microwave measurements confirmed measurements of earlier ER-2 flights of the direct anticorrelation of ClO. MLS added information on the global extent of the presence of high concentrations of ClO in the high-latitude lower stratosphere in spring, where the ozone hole formed (Figure 5.3; Waters et al. 1993). Thus, the ozone values had decreased at the same time and locations where high values of ClO occurred over Antarctica in the southern spring.

More recently, spaceborne measurements of bromine oxide (BrO) have been made by instruments measuring reflected UV-visible radiation (McElroy et al. 1986, Tung et al. 1986, Sinnhuber et al. 2005) and microwave emissions (Livesey et al. 2006). BrO is even more effective in the photolytical destruction of ozone; 50 times more effective than ClO on a molecule per molecule basis. The satellite data are broadly consistent with current understanding of bromine chemistry, indicating that at the observed concentrations BrO plays a significant role in the budget of lower-stratospheric ozone. Taken together, these and related data on other species confirmed the chemistry in the coupled models of the stratosphere, greatly improving their utility and trustworthiness as tools to guide policymakers.

Although satellite instruments did not discover the severely disturbed ozone conditions in southern polar regions, satellite observations from the BUV series of instruments provided unique detailed maps of the Antarctic ozone hole (Figure 5.4). The monthly mean ozone column over Antarctica provides information on the evolution of the Antarctic ozone hole from the first measurements in 1970 until 2005. These maps allowed tracking of its size and depth every year, providing the most extensive information on its annual growth, extent, and decay crucial to ozone assessments and to the amendments to the Montreal Protocol (WMO 2006). In the first measurements a crescent of higher ozone can be observed, generally centered south of Australia, with a lower amount over Antarctica itself. With the passage of time, the

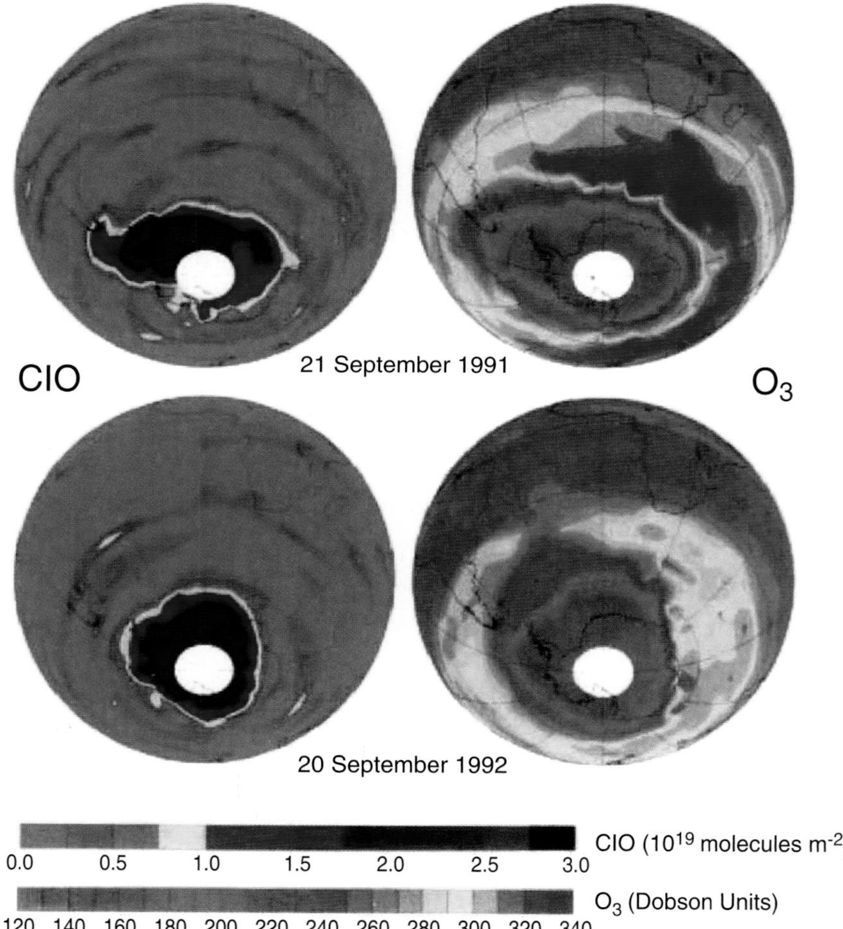

FIGURE 5.3 Chlorine monoxide (ClO; left panel) and stratospheric ozone (O_3; right panel) columns over the southern hemisphere measured by the Microwave Limb Sounder (MLS) on the Upper Atmosphere Research Satellite (UARS) for days during the austral springs of 1991 and 1992. These images show that high ClO concentrations coincide in space and time with low O_3 concentrations confirming ground-based measurements and the proposed mechanisms for ozone depletion. The white circle over the pole indicates area where no data is available. SOURCE: Waters et al. (1993). Reprinted with permission from Macmillian Publishers Ltd., copyright 1993.

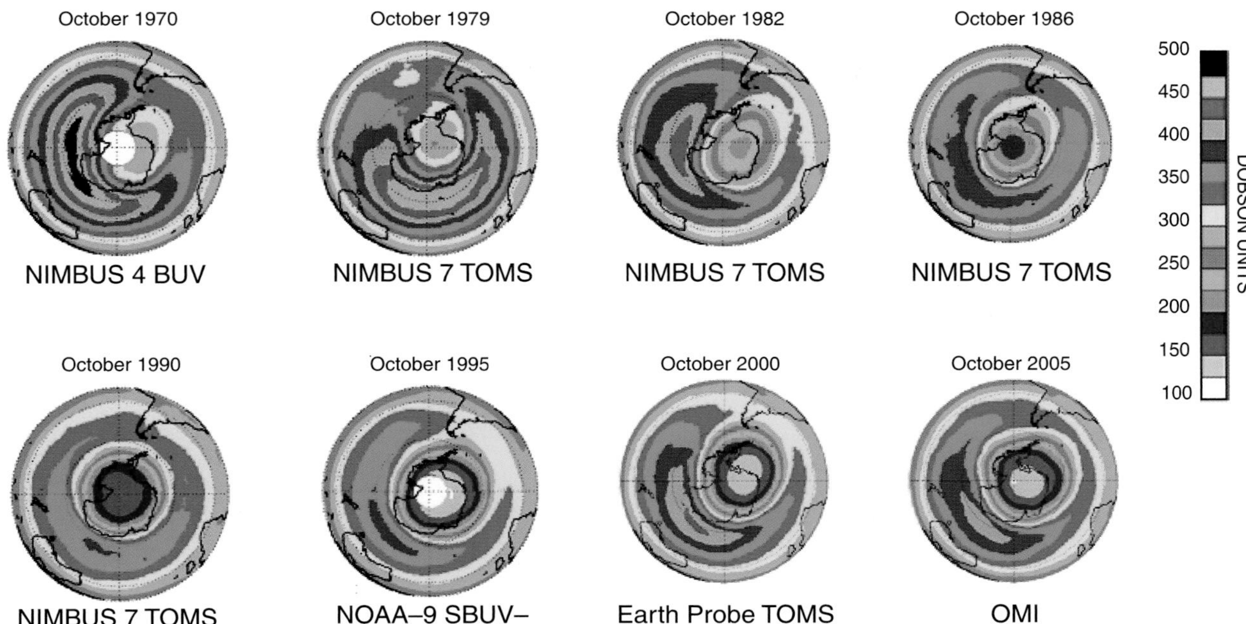

FIGURE 5.4 October monthly mean total ozone column over the southern hemisphere for 8 selected years between 1970 and 2005. These show large interannual variations, with the hole generally becoming larger and deeper until recent years. SOURCE: Data provided by R. McPeters, NASA GSFC; modified by John Gille.

amount over Antarctica decreased sharply. The lowest ozone values were observed in 1995, with a slight increase since then. Analysis of the area and extent of the ozone hole for the years 2005 and 2006 compared to the mean from 1979 to 2005 indicates that the extent of ozone depletion over Antarctica is greater during austral spring in the most recent years compared to the mean (Figure 5.5). The maximum extent is usually reached near the end of September.

OZONE DEPLETION OVER THE NORTHERN HEMISPHERE

Although the Antarctic ozone hole is the better-known phenomenon that has illuminated and confirmed the theory of halogen-catalyzed chemical ozone destruction, a more important question from a societal point of view is the effect of anthropogenic chlorine, bromine, and other gases on the ozone concentrations over heavily populated nonpolar latitudes. Because ground-based measurements could never provide the necessary coverage or sampling frequency, satellite measurements are essential in determining the extent of global ozone. The measurements of globally averaged ozone are sufficiently stable and precise to be able to detect a 3 percent decrease in the northern hemisphere midlatitudes from 1979 to 1997 and a 6 percent decrease in the southern hemisphere over the same period.

Since 1997, in response to international regulations, concentrations of chlorine-containing gases in the atmosphere have decreased and the rate of depletion of stratospheric ozone has slowed (WMO-UNEP 2006). Data in that report provide some indications of the beginning of a recovery (Yang et al. 2006; Figure 5.6). To facilitate the search for trends due to halogen-induced destruction, variations due to seasonal, solar, and quasi-biennial effects have been removed from the ozone time series. These corrected ozone values, shown in Figure 5.6, display a decrease from 1979 until the mid-1990s, after which they seem to increase.

In conclusion, satellite observations provided the first measurements of the global vertical, horizontal, and temporal distributions of ozone and dynamical variables in the stratosphere, permitting the monitoring of their long-term changes. By allowing verification of the factors causing those changes, satellite observations were critical in confirming the seriousness of the danger posed by the release of anthropogenic halocarbons and, thus, in leading to the international agreement to protect the ozone layer. "The evolution of scientific understanding of ozone depletion and related policy decisions has since been heralded as one of the most remarkable environmental success stories of the 20th century" (NRC 2007b). It has created conditions for the recovery of the ozone layer to preindustrial conditions and removed a major hazard to human health and the biosphere.

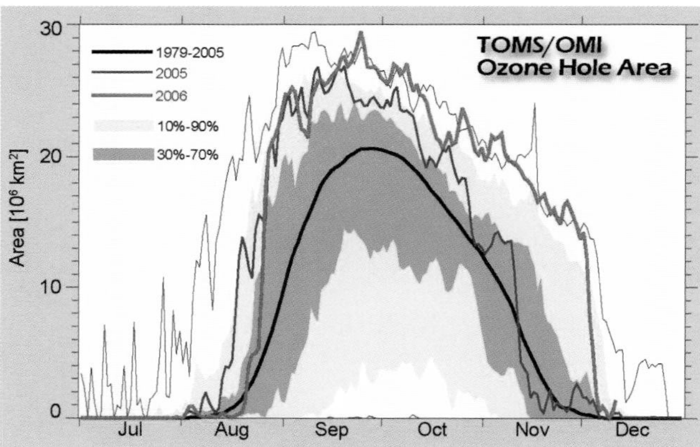

FIGURE 5.5 The area of the ozone hole with less than 220 dobson units as deduced from TOMS and the Ozone Monitoring Instrument (OMI) data for 2006 (red) and 2005 (blue). The thick black line indicates the 1979-2005 mean, with the light-blue area giving the 10th-90th percentiles over that period and the blue-green area giving the 30th-70th percentiles. The thin gray line shows the maximum over this period. The area in 2006 was occasionally the highest on record. SOURCE: WMO (2006). World Meteorological Organization, copyright 2006.

TROPOSPHERIC OZONE AND TRACE GASES

The troposphere presents special challenges to passive satellite detection because of clouds and diminished viewing by limb or occultation methods in the lower atmosphere. Nonetheless, space-based studies of tropospheric ozone began in the mid-1980s, and other critical trace species became measurable from 1995 onward with the launch of the European Space Agency's (ESA) Global Ozone Monitoring Experiment (GOME; later SCIAMACHY, GOME II) and a constellation of the National Aeronautics and Space Administration's (NASA) Earth Observing System (EOS) and related satellite instruments (Table 5.1). The chemical species listed are those of sufficient concentration, lifetime, and distribution to be detectable from space. Given the lifetimes and interactions of tropospheric gases and aerosols (discussed in Chapter 4), the variability of these species with synoptic processes reveal as much about tropospheric transport as chemistry. Furthermore, a few other processes (e.g., lightning, which is detected from space) are proxies from which trace gas concentrations are often inferred.

Tropospheric Ozone in the Tropics: "First Success"

Tropical tropospheric ozone deserves special mention because it has been derived from instruments designed to measure total and stratospheric ozone and because its time series, dating from the Nimbus era (above) is sufficiently long that trends and climate signals are detectable in the satellite record. Differencing the total and the stratospheric ozone column amount to deduce a "tropospheric residual" is possible due to the observation that stratospheric ozone is zonally invariant in the tropics and changes slowly over week-to-month timescales. Thus, a stratospheric profiler like (SAGE, later SBUV, HALOE, and MLS) is used to determine the stratospheric ozone column to be subtracted from a BUV-based total ozone instrument such as TOMS or OMI.

The first maps of the so-called tropospheric ozone resid-

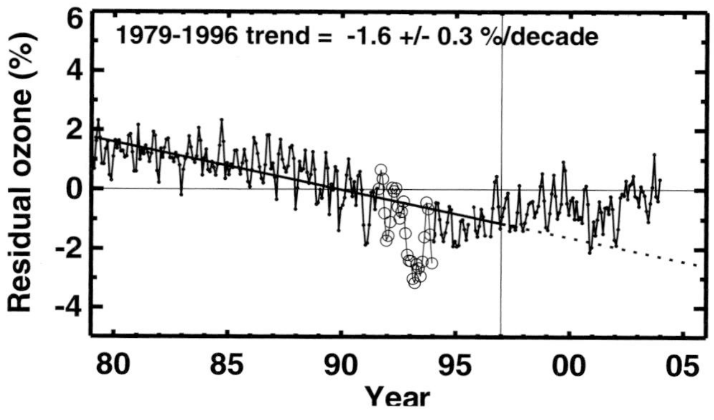

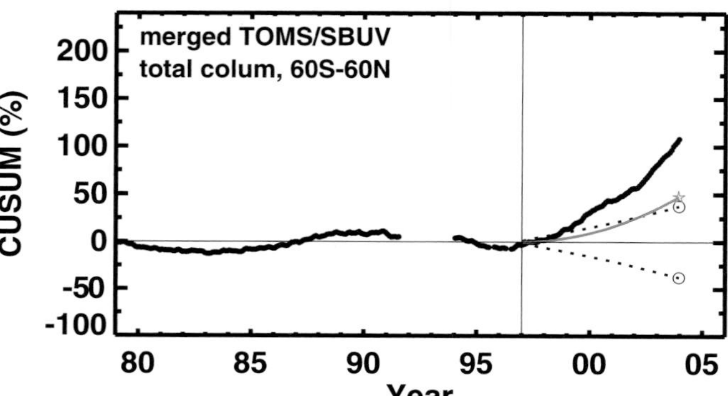

FIGURE 5.6 Left panel: Time series of monthly average ozone based on merged TOMS-SBUV total columns between 60° S and 60° N for 1979-2005, with effects due to other causes removed. The trend line indicates the ozone trend calculated from the data for 1979-1996 (solid line) and projected linearly thereafter. Right panel: Cumulative sum of differences from the mean trend line (percent). The solid straight line indicates the line fitted to the ozone trend calculated from the data for 1979-1996 and projected linearly thereafter (dotted line). The solid black line rising above the green line is an indication that ozone recovery has begun. SOURCE: Yang et al. (2006). Reprinted with permission by American Geophysical Union, copyright 2006.

TABLE 5.1 Space-Based Studies of Tropospheric Ozone and Other Critical Trace Species

Chemical Species	Instrument/Technique	References
Tropospheric O_3	TOMS-SAGE differencing	Fishman et al. (1991)
	TOMS-SBUV differencing	Fishman et al. (1996, 2005)
	OMI-HALOE differencing (on Upper Atmosphere Research Satellite [UARS])	Ziemke et al. (1998)
	TOMS/OMI-MLS differencing	Ziemke et al. (1998) Chandra et al. (2003)
	"Cloud Slicing"	Ziemke et al. (2001) Ziemke and Chandra (2005)
	OMI Assimilation	Stajner et al. (2006) Pierce et al. (2007)
	TES (Tropospheric Emission Spectrometer)	Worden et al. (2007)
CO	TES	Rinsland et al. (2006)
	MOPITT (Measurement of Pollution in the Troposphere, on Terra Satellite)	Lamarque et al. (2003) Deeter et al. (2003) Yurganov et al. (2004)
	AIRS (Atmospheric Infrared Sounder)	McMillan et al. (2005)
NO_2 SO_2 BrO HCHO	GOME (I, 1995; II, 2005) and SCIAMACHY (SCanning Imaging Absorption spectroMeter for Atmospheric CHartographY)	Richter and Burrows (2002) Eisinger and Burrows (1998) Carn et al. (2005) Richter et al. (1998) Hollwedel et al. (2004) Chance et al. (2000)

ual were seasonal averages in the tropics and subtropics that revealed a distinctive zonal wave-one pattern in tropospheric ozone in the southern hemisphere (Fishman and Larsen 1987, Fishman et al. 1991, 2003; Figure 5.7). The minimum in the ozone residual occurs in the central-western Pacific where photochemical sources are few and convection associated with the Walker circulation maintains a low-ozone column throughout the troposphere (Kley et al. 1996, Thompson et al. 2003). The South Atlantic maximum is characterized by a tropospheric ozone column with an amplitude approximately 10-15 Dobson units (DU) greater than that over the Pacific.

Interestingly, the Atlantic tropospheric ozone maximum is largest at the end of the southern hemisphere biomass burning season, from August through November. A fire-ozone linkage was established through the 1992 Southern Africa Fire-Atmosphere Research Initiative Transport and Atmospheric Chemistry near the Equator-Atlantic ground, multiaircraft and balloon campaigns (van Wilgen et al. 1997), using ozone, ozone precursor, and free radical measurements over South America and southern Africa (Fishman et al. 1996, Jacob et al. 1996). The synergism of satellite and in situ measurements in these experiments, with aircraft flying toward satellite-observed ozone maxima, ushered in a new era for tropospheric chemistry—just as 5 years earlier, airborne ozone depletion missions targeted regions where the TOMS satellite pinpointed column ozone loss.

Further studies with satellites have shown the South Atlantic ozone maximum to be more complex than initially assumed. First, the late burning season overlaps the start of the tropical rains, suggesting that biogenic nitrogen oxide from wet soils (Harris et al. 1996) and lightning nitrogen oxide (Moxim and Levy 2000) also contribute to the ozone burden in September and October. The location and amount of lightning have been observable only with Optical Transient Detector on MicroLab-1 and the Tropical Rainfall Measuring Mission (TRMM) Lightning Imaging Sensor (Christian et al. 1989, Bocippio et al. 2000). Second, closer inspection of tropospheric ozone maps showed the South Atlantic maximum was year-round, exemplified by the so-called tropical ozone paradox, named for the persistence of the maximum in January and February when biomass burning was a maximum north of the Intertropical Convergence Zone (Thompson et al. 2000). The causes of the paradox were analyzed with sondes (Jenkins et al. 2003, Chatfield et al. 2004, Jenkins and Ryu 2004), aircraft data (Sauvage et al. 2006), and other satellites, principally the Measurements of Pollution in the Troposphere (MOPITT; Edwards et al. 2003).

Tropospheric Views Since 1995

Breakthroughs in our understanding of tropospheric composition escalated after the 1995 launch of GOME, with its 2002 follow-on mission, the Scanning Imaging

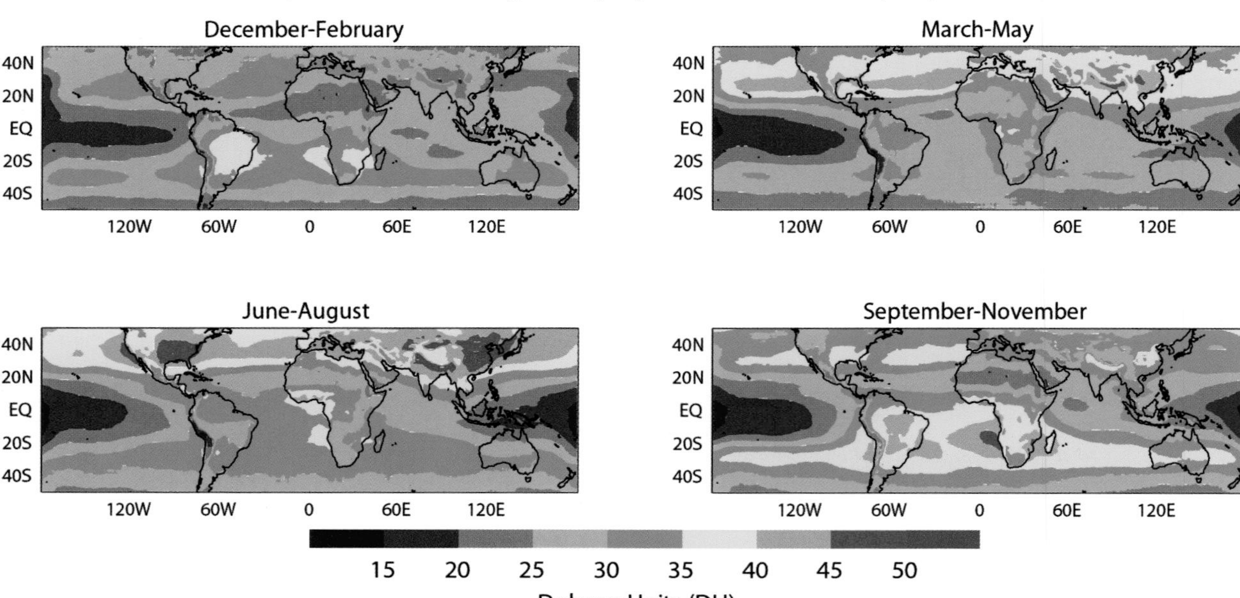

FIGURE 5.7 Seasonally averaged tropospheric ozone column, so-called residual amounts, that show high ozone in northern midlatitude spring and during the late biomass burning season over South America and Africa. SOURCE: After Fishman et al. (2003). Reprinted with permission from the European Geosciences Union, copyright 2003.

Absorption Spectrometer for Atmospheric Chartography (SCIAMACHY), and with the EOS constellation of satellites (1999, 2002, 2004), each of which has instruments sensing lower-atmospheric trace gases, aerosols, and clouds. Multiple methods and sensors have been used to measure most of these constituents (Table 5.1).

Tropical Ozone

Refined satellite products, including several using the ozone residual concept, showed greater complexity in tropospheric ozone, notably in the tropics. During the 1997-1998 El Niño-Southern Oscillation (ENSO), upper-tropospheric ozone increased and water vapor decreased due to enhanced subsidence from the lower stratosphere (Chandra et al. 1998). A time series of tropospheric ozone derived from TOMS back to 1980 showed signatures of ENSO events in the 1980s (Thompson et al. 2001). Pollution from the Indonesian fires, instigated by the 1997-1998 ENSO drought, created tropospheric ozone that TOMS followed across the Indian Ocean. New aerosol products (from TOMS and the Sea-Viewing Wide Field-of-View Sensor [SeaWiFS]) project tracked pollution day to day, showing that during the worst health episodes, smoke and ozone were decoupled (Thompson et al. 2001; Figure 5.8).

A dedicated tropical ozonesonde validation network for satellite instruments has pinpointed ozone interactions with dynamics (Thompson et al. 2003). Profiles from soundings combined with the global view afforded by SAGE, HALOE, MLS, and Atmospheric Infrared Sounder (AIRS) instruments, have characterized the natural modulation of water vapor and ozone in the tropical tropopause region. This has fostered the growth of a subdiscipline of "tropical tropopause layer" studies, including cirrus clouds as well as trace gases (Gettelman et al. 2002, Folkins and Martin 2005, Dessler and Minschwaner 2007, Takashima and Shiotani 2007).

Other Trace Gases in the Troposphere

The reach of pyrogenic pollution is sometimes surprising. The air quality community has used satellite measurements of cardon monoxide (CO), ozone, and smoke to discriminate local and imported pollution for regulatory purposes (Morris et al. 2006, Pierce et al. 2007), especially in the case of boreal fires. An important feature of satellite CO instrumentation is that detection is strongest within midtropospheric layers where the gas has been introduced by convection. Indirectly, then, regions of maximum convective activity are identified through chemical measurement.

The power of spaceborne CO measurements was proven with the Measurement of Air Pollution from Satellites (MAPS) Shuttle instrument (1984-1994; Connors et al. 1999) but only since MOPITT was launched on the Terra platform have global observations of this key "ozone precur-

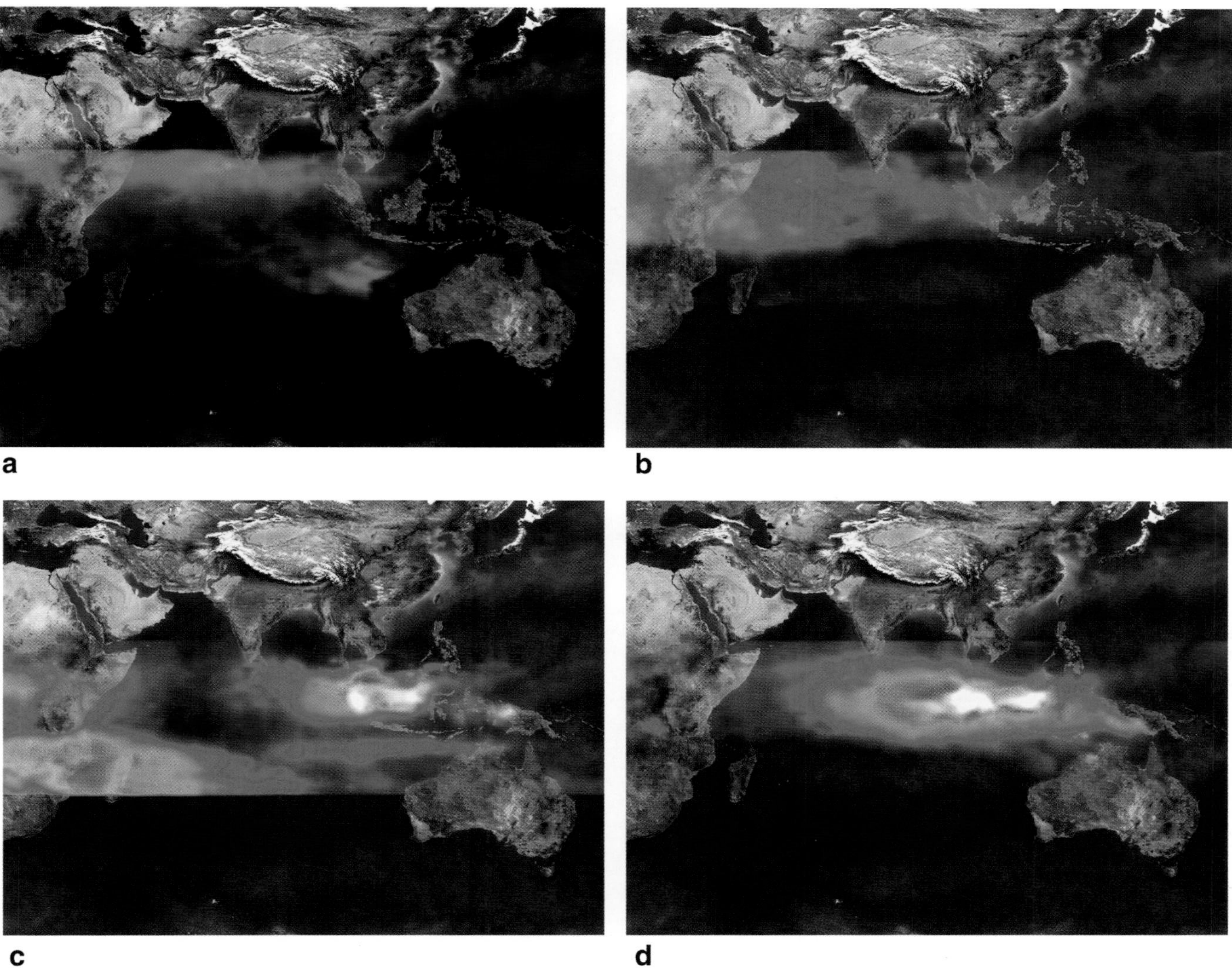

FIGURE 5.8 Tracking pollution using data from NASA's TOMS satellite instrument. In 1997 smoke from Indonesian fires remained stagnant over Southeast Asia while smog (tropospheric, low-level ozone) spread more rapidly across the Indian Ocean toward India. This situation was exacerbated by ENSO, which had already increased the thickness of smog over the region. At the same time, additional smog from African fires streamed over the Indian Ocean and combined with the smog from Indonesia in mid-October (lower right), creating an aerial canopy of pollutants. SOURCE: NASA.

sor" constituent been available (Figure 5.9). Ozone in the free troposphere has a lifetime of weeks to a month or more; for CO the photochemical lifetime is several months. MOPITT CO shows transhemispheric transport properties similar to ozone in the "paradox" region of the South Atlantic (Edwards et al. 2003). An AIRS product (McMillan et al. 2005) also tracks CO from industrial activity, and boreal and tropical fires over thousands of kilometers. Transboundary and transoceanic pollution among industrialized regions shows expected patterns.

Nitrogen oxides[1] (NO_x) are released by combustion along with carbon monoxide. However, the chemical NO_x lifetime is much shorter (hours), so sources are readily identified. GOME and OMI NO_2 appears most intense in industrial regions compared to biomass burning, but it shows up during the tropical rainy season when soil release is expected to make a significant contribution (Jaeglé et al. 2004). Models must be used to infer NO_x from lightning, and the conversion

[1]More specifically, nitrogen dioxide (NO_2) in equilibrium with the prime emittant nitric oxide (NO). The sum, NO + NO_2, is designated as NO_x.

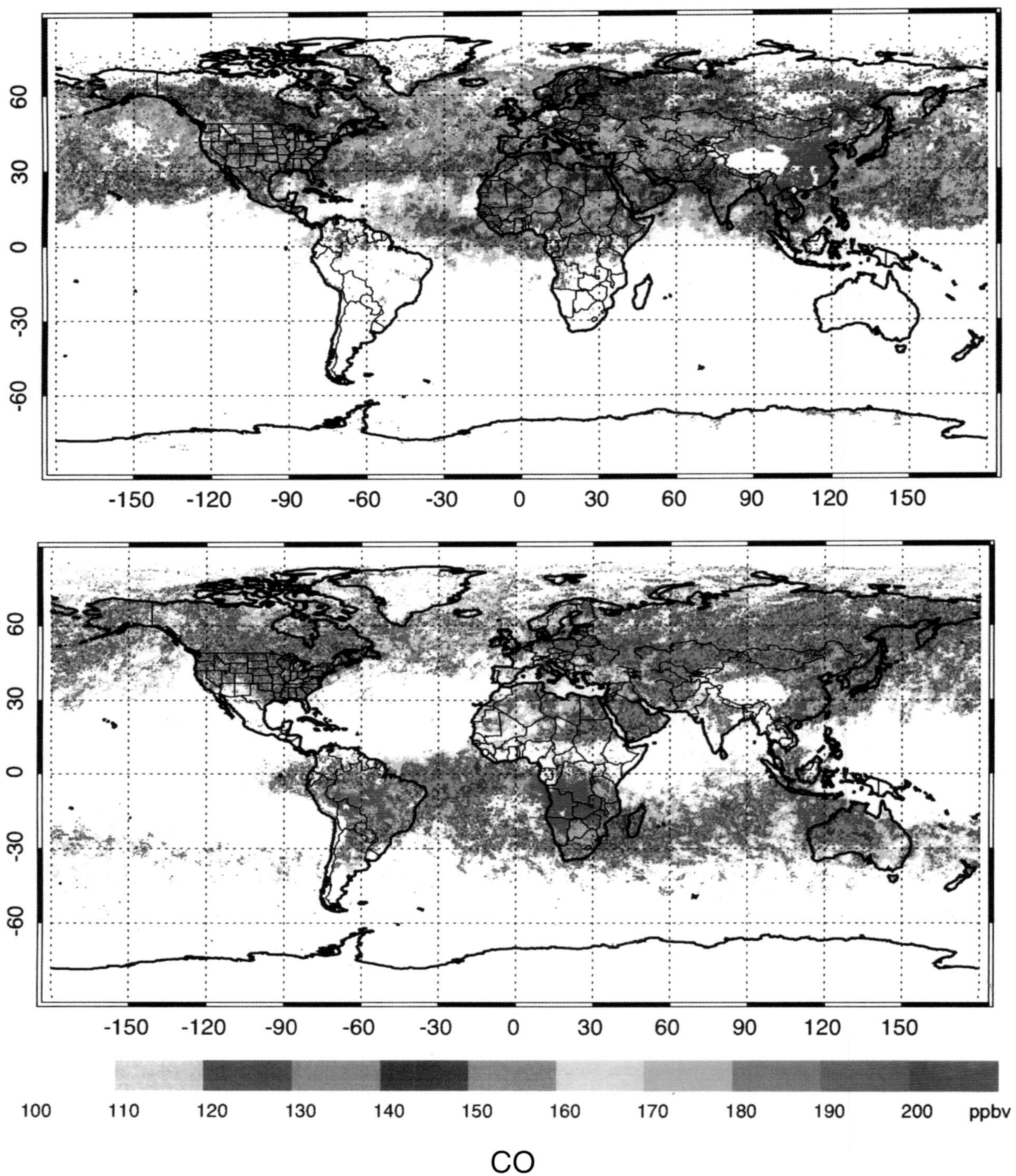

FIGURE 5.9 The seasonally changing global distribution of CO pollution observed by Terra MOPITT at an altitude of 700 hPa (about 3 km). Averages are shown separately for March 2006 (top) and September 2006 (bottom). High CO pollution levels are shown in red. In addition to chemical production, northern hemisphere pollution sources are predominantly urban and industrial, while high CO in the tropics and southern hemisphere often results from biomass burning. NOTE: ppbv = parts per billion by volume. SOURCE: Modified from Edwards et al. (2006). Reprinted with permission by American Geophysical Union, copyright 2006.

of lightning flashes to tropospheric NO_2 release has been parameterized in several ways.

Satellites have allowed mapping of other important tropospheric trace gases and have been essential in solving the mystery of "polar sunrise" tropospheric ozone depletion. Since the 1980s, Arctic ozone has been known to disappear at the surface in the first few weeks of spring (Barrie et al. 1988). Organic halogen in some form was originally implicated, but the mechanisms were unclear until GOME detection of BrO (Richter et al. 1998, Hollwedel et al. 2004) in the first sunlit days (Figure 5.10). Reactions with highly saline surface associated with annual sea ice are now believed to be the source of airborne labile halogen compounds, which cause the surface ozone depletion (Rankin et al. 2002). The same phenomenon is detected at the edge of the Antarctic continent in austral spring.

For many years satellite measurements of stratospheric composition have advanced our understanding of the chemistry and dynamics of the region above the tropopause. As this region continues to respond to changes in halocarbon concentrations and global temperature, the measurements will continue to be vital to monitoring the health of the planet. Furthermore, the present growth of greenhouse gases leads not only to warming of the troposphere but also to cooling of the stratosphere, which is predicted to affect the rate and extent of ozone recovery. Continuation of the types of measurements described above is essential to monitoring the progress of ozone recovery and to further the understanding of the complex role of ozone in the climate system.

Although satellite measurements of tropospheric species are more difficult, rapid advances in measurements of tropospheric composition are providing insights into the sources, mechanisms, and transport of many species. Combined with data assimilation schemes, continuing tropospheric chemistry observations from satellites will lead to a better understanding of the factors affecting air quality and the ability to predict its interactions with the stratosphere and climate system.

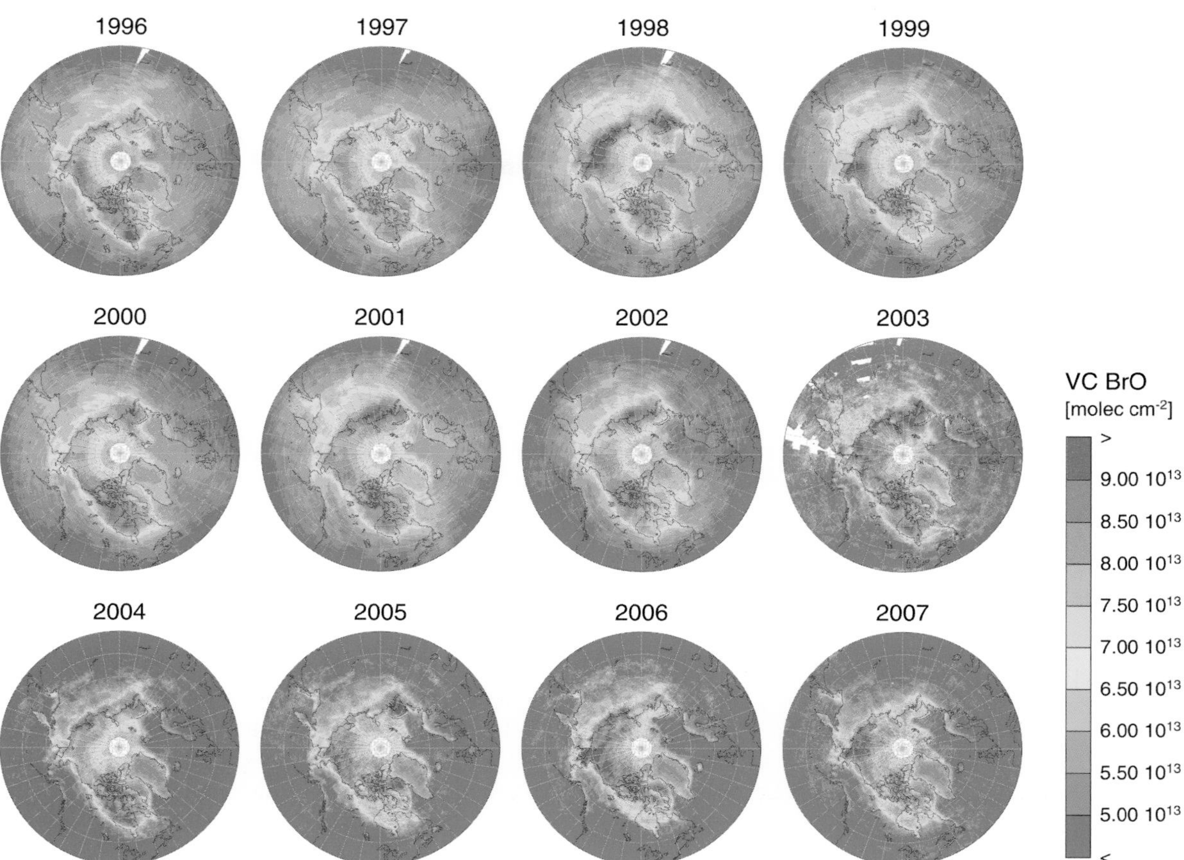

FIGURE 5.10 BrO from GOME (April monthly averages, 1996-2002) SCIAMACHY (2003-2007). Total column BrO includes more or less uniform stratospheric and free tropospheric contributions. The majority signal is from boundary layer BrO that forms from heterogeneous processes associated with the annual sea ice (Richter et al. 1998, Hollwedel et al. 2004). SOURCE: Figure courtesy of A. Richter and J.P. Burrows, University of Bremen, Germany.

6

Hydrology

The global water cycle links all components of the Earth system as water moves from the atmosphere to the land and ocean and through the biosphere and cryosphere. Water plays a central role in the climate system due to its importance in the carbon cycle as water availability is a key factor for terrestrial photosynthesis. Water is also essential to the energy balance of the Earth because the rate of evaporation controls the latent heat flux. In addition, water is necessary to sustain all life; and, consequently, furthering the understanding of global hydrology is of central importance to society.

Through some remarkable technological accomplishments, some important variables and processes associated with the water cycle can be retrieved now from satellite observations: water vapor (see Chapters 3 and 5), precipitation over oceans and land, snow, ice sheet mass and flow (see Chapter 7), continental groundwater storage, and sea surface temperature (see Chapter 8). The new perspective provided by satellite observations—revealing high temporal and spatial variability—has transformed the global understanding of hydrology more than any single observing platform could have done. At the same time, some important hydrologic measurements are not yet available from space, such as snow water equivalent in mountainous areas, soil moisture, and procedures to estimate evapotranspiration from remote sensing.

PRECIPITATION ESTIMATES FROM THE TROPICAL RAINFALL MEASURING MISSION

In orbit since 1997, the Tropical Rainfall Measuring Mission (TRMM) has transformed our ability to measure the spatial and temporal variability of rainfall in the tropics, especially over the oceans. Learning about precipitation over the oceans has catalyzed further understanding of air-sea interaction, the role of runoff to the seas in ocean circulation, and the vertical circulation of the oceans. In addition, it has led to great improvements in weather forecast skill, particularly for the southern hemisphere (see Chapter 3). TRMM's ability to measure the height in the atmosphere where precipitation is generated provides information on the vertical distribution of release of latent heat, which in turn improves our knowledge of atmospheric circulation and climate. Moreover, TRMM has demonstrated that the technology for reliable precipitation measurements from space is now available and has provided the community with important lessons to guide the design of the Global Precipitation Mission (GPM; NRC 2007c).

Before TRMM, rainfall estimates were obtained from ground-based sources (e.g., rain gauges and radar) and from satellites with visible, infrared, and passive microwave sensors (e.g., Advanced Microwave Scanning Radiometer [AMSR] for the Earth Observing System [EOS], Advanced Microwave Sounding Unit B, and Special Sensor Microwave/Imager [SSM/I]). The scientific accomplishments of TRMM are due primarily to two innovative aspects of the mission: TRMM's complementary suite of instruments and its orbital characteristics (NRC 2006).

TRMM's instruments include a microwave imager, a visible and infrared scanner, and a lightning imaging sensor all on the same platform along with the first-ever precipitation radar in space. The suite of instruments on TRMM allows for intercalibration among the instruments as well as cross-calibration with sensors on other platforms. TRMM's precipitation radar provides direct, fine-scale observations of precipitation and its vertical distribution. The satellite's 35-degree inclination orbit and low altitude (402.5 km) allows for sampling well beyond the tropics to 60° N/S, but sampling in the tropics is more frequent. The orbit is not sun-synchronous; therefore, in each month it acquires measurements at all longitudes and all times of day. These advantages augment the spatial and temporal views of standard polar-orbiting environmental satellite trajectories. However, due to TRMM's narrow swath, data for any given storm or location are available infrequently. The GPM proposes to overcome most of TRMM's limitations and is central to ensuring the availability of remotely sensed precipitation measurements for climate research (NRC 2007a).

Through its technological innovations, TRMM has enabled the following scientific accomplishments for hydrology and climate: establishing rainfall climatology, quantifying the diurnal cycle of precipitation and convective intensity, and profiling latent heating (NRC 2006). TRMM data have also contributed to operational use: near-real-time TRMM-based multisatellite estimations of rainfall are being used to detect floods in the United States and especially overseas where conventional information is lacking. The National Oceanic and Atmospheric Administration's (NOAA) National Environmental Satellite Data and Information Service uses TRMM data as part of its Tropical Rainfall Potential Program to estimate flood potential in hurricanes (Box 6.1, Figure 6.1). The National Aeronautics and Space Administration's (NASA) TRMM-based Multisatellite Precipitation Analysis is used globally to detect floods and monitor rain for agricultural uses. The Naval Research Laboratory Monterey and the National Centers for Environmental Prediction use TRMM data as a key part of their multisatellite rain estimates. TRMM data are central to the success of these efforts because of their accuracy and the significant sampling coverage by TRMM in the tropics.

The scientific accomplishments and operational advantages of TRMM have spurred the development of the GPM follow-on mission, scheduled for launch in 2013 (NRC 2007a). GPM will consist of a core spacecraft with a dual-frequency precipitation radar and a multifrequency microwave radiometric imager with high-frequency capabilities to serve as an orbiting "precipitation physics laboratory." In addition to the core spacecraft, GPM will include a constellation of current and planned satellites with passive microwave radiometers. Together, the system will provide calibrated global precipitation at 2- to 4-hr intervals.

SEASONAL SNOW COVER

Of the seasonal changes that occur on Earth's land surface, perhaps the most profound is the accumulation and melt of seasonal snow cover. Snow influences climate, weather, and the water balance. Snow cover has significant effects on energy and mass exchange between Earth's surface and atmosphere and is an important reservoir of fresh water. Its high albedo changes the surface radiation balance; its low thermal diffusivity insulates the ground; and it is a wet, cold surface in the context of heat and moisture fluxes. Therefore, snow cover exerts a huge influence on the hydrologic cycle during the winter and spring for much of Earth's land area. Near many mountain ranges, the seasonal snow cover is the dominating source of runoff, filling rivers and recharging aquifers that more than a billion people depend on for their water resources (Barnett et al. 2005a). Snow affects large-scale atmospheric circulation. Early-season snow cover variability in the northern hemisphere, for example, leads to altered circulation patterns, suggesting implications for climate predictability (Cohen and Entekhabi 1999).

For four decades, satellite remote sensing instruments have measured snow properties. These weekly measurements represent one of the longest satellite-derived climate data records, which now enables scientists to study long-term trends in seasonal snow cover (Frei and Robinson 1999). At optical wavelengths, sensors such as the NOAA Advanced Very High Resolution Radiometer (AVHRR) and the Landsat Thematic Mapper (TM) have been used to produce maps of snow cover at both continental and drainage-basin scales. In the EOS era, snow-cover products are available from the Moderate Resolution Imaging Spectroradiometer (MODIS), the Multiangle Imaging Spectroradimeter (MISR), and the Advanced Spaceborne Thermal Emission Reflection and Radiometer (ASTER). Snow-water equivalent (the depth of liquid water that the snowpack would produce if it melted) is regularly estimated at coarse spatial resolution from passive microwave data, including SSM/R, SSM/I, and the EOS instrument AMSR-E in a time series that goes back to 1978. However, at finer spatial resolution, necessary for the mountains, measuring snow-water equivalent is a difficult problem; and a proposed sensor for Snow and Cold Land Processes (SCLP) is recommended as one of 17 high-priority missions for launch before 2020 (NRC 2007a).

König et al. (2001) and Dozier and Painter (2004) have reviewed developments in remote sensing of snow and ice. Among them is the use of snow-covered area from MODIS in hydrologic analysis and modeling (Box 6.2, Figure 6.2). Through updates of a runoff model with measurements of snow cover, seasonal streamflow forecasts have been improved (McGuire et al. 2006). Unlike surface measurements, satellite observations are able to show the distribution of snow over the topography, revealing that considerable snow at higher elevations remains after all snow has disappeared from the surface measurement stations.

An additional property measured from MODIS is snow albedo. In the current generation of climate and snowmelt models, snow albedo is typically either prescribed or represented by empirical aging functions, when truly it is a dynamic variable affected by grain growth and light-absorbing impurities. Newer analyses of snow cover are incorporating the seasonal evolution of both the snow cover and its albedo. In the visible part of the spectrum, clean, deep snow is bright and white, irrespective of the size of the grains. Beyond the visible wavelengths in the near infrared and shortwave infrared, however, snow is one of the most "colorful" substances in nature. Newly fallen snow usually has a fine grain size, but metamorphism and sintering throughout the winter and spring increase the grain size, bond grains together, and reduce reflectance in wavelengths beyond about 0.8 μm (Warren 1982). This behavior of snow is important to the snowpack's energy balance because the decrease in albedo often occurs during the spring when the

BOX 6.1
Improved Understanding of Hydrology and Climate from TRMM

TRMM-based multisatellite data are being used as input into hydrologic models, including the Land Data Assimilation System, to better understand land-atmosphere interactions on scales of days to years (Rodell et al. 2004) and to study variations in river runoff (Fekete et al. 2004). These same data are being used to monitor crops in Central America and elsewhere and as input into river forecast models in South Asia and other locations. Analysis of TRMM precipitation radar data has been used to discover orographic precipitation processes and diurnal cycles of rainfall causing flash floods in headwater streams (Barros et al. 2004). TRMM observations led to the discovery of extremely tall convective towers within the vertical precipitation profiles of tropical cyclones. Kelley et al. (2004) reported that the chance of intensification increases when one or more of these "hot towers" exist in the tropical cyclone's eyewall (Figure 6.1).

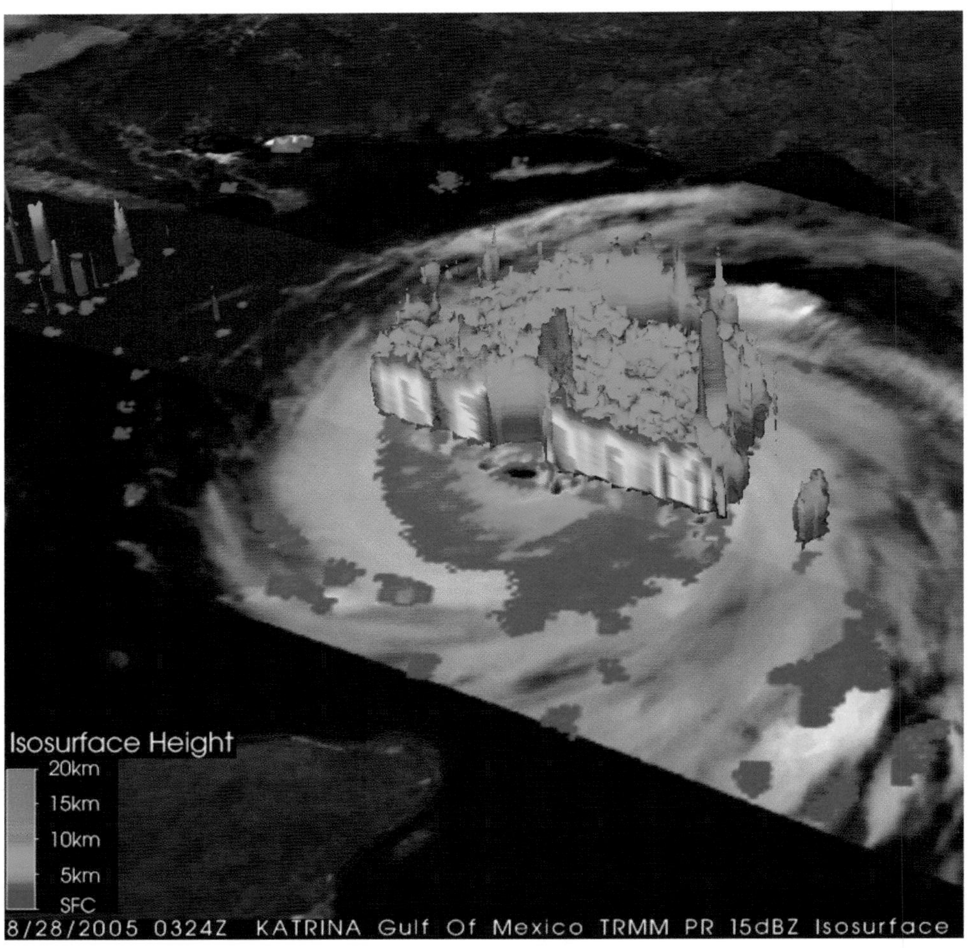

FIGURE 6.1 A cross-sectional view of Hurricane Katrina through the eye of the storm, as observed from TRMM. This image shows the horizontal distribution of rain intensity on August 28, 2005, when Katrina was a Category 3 hurricane with maximum sustained winds of 100 knots (115 mph). Rain rates in the central portion of the swath are from the TRMM precipitation radar, and the rain rates in the outer swath are from the TRMM microwave imager. The rain rates are overlaid on infrared data from the TRMM visible infrared scanner. Two isolated hot towers (in red) are visible: one in an outer rain band and the other in the northeastern part of the eyewall. The height of the eyewall tower is 16 km. Towers of this height near the core are often an indication of intensification as was true with Katrina, which became a Category 4 storm soon after this image was taken. SOURCE: NASA (2005).

BOX 6.2
High-Resolution Seasonal Snow Cover Data Improve Climate and Hydrology Models

Because of the influence of seasonal snow cover on climate, weather, and water balance, it is a crucial quantity for climate and hydrology models. Furthermore, daily maps are necessary for hydrologic and climate models due to the dynamic nature of snow cover, which changes at a slower timescale than atmospheric phenomena but faster than other surface covers. The availability of daily global observations of this parameter was inconceivable prior to the satellite era. Nowadays, the global MODIS snow-cover product (Hall et al. 2002) is produced daily and as an 8-day composite at 500-m spatial resolution. For global climate models, daily snow cover is produced at 0.05° latitude-longitude grid cells (about 5.5 km in the north-south direction) along with monthly global composites. The composites are necessary because cloud cover and viewing geometry affect the daily images (Figure 6.2).

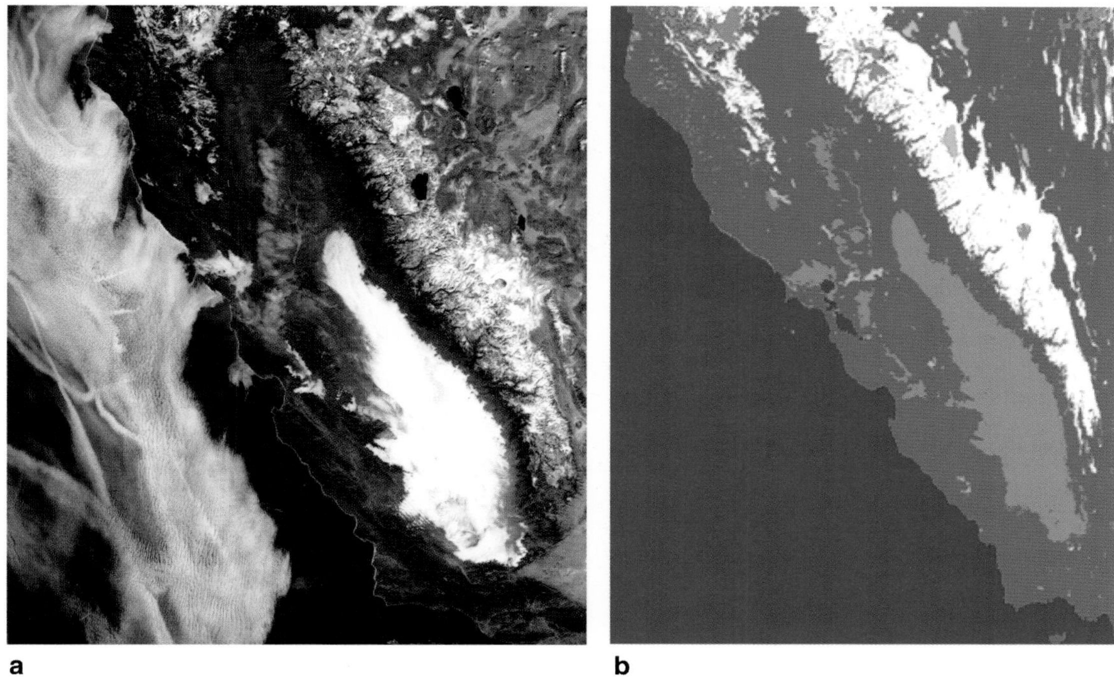

FIGURE 6.2 MODIS image (left) and interpreted snow (white) and cloud (pink) cover over the Sierra Nevada, January 5, 2003. SOURCE: *http://modis-snow-ice.gsfc.nasa.gov/images.html.*

incoming solar radiation becomes greater as the solar elevation increases and the days get longer.

In the context of hydrologic models, this albedo decay has spatial variability. Molotch et al. (2004) examined snow ablation from a grid-based distributed snowmelt model, using field data from extensive snow surveys during the melt season to initialize the model with a spatial distribution of snow-water equivalent and then to test the model with subsequent surveys. Remotely sensed albedo typically differed by 20 percent from albedo estimated using a common snow age-based empirical relation applied uniformly across the domain. Snowpack models are just beginning to incorporate albedo evolution, based on the movement of water molecules in the snow to reduce the surface area of the grains in comparison to their volume (Flanner and Zender 2006).

A recent development in mapping snow cover and its albedo is "subpixel" analysis. Snow-covered area in mountainous terrain usually varies at a spatial scale finer than that of the ground instantaneous field of view of the remote sensing instrument. This spatial heterogeneity poses a "mixed-pixel" problem because the sensor may measure radiance reflected from snow, rock, soil, and vegetation. To use the snow characteristics in hydrologic models, snow must be mapped at subpixel resolution in order to accurately represent its spatial distribution; otherwise, systematic errors may result. For example, especially in drier years, much of the snow cover is patchy at the lower elevations. An image classification that identifies each pixel as either snow covered or not may miss much of this snow.

Mapping of surface constituents at subpixel scale uses a technique called "spectral mixture analysis," based on the assumption that the radiance measured at the sensor is a linear combination of radiances reflected from individual surfaces (Figure 6.3). Snow does not have a unique reflectance in each wavelength band, but given its physical characteristics such as grain size and amount and composition of impurities, a snow end member can be chosen that results in the lowest error in the solution of the simultaneous equations (Painter et al. 2003). The information thereby derived is the fractional snow-covered area for each pixel and the albedo of that snow.

DISCOVERY OF ANCIENT BURIED RIVER CHANNELS

In 1981 the first Shuttle Imaging Radar (SIR-A) was launched on the space shuttle Columbia, assembled partly with spare parts from the 1978 Seasat synthetic aperture radar (SAR). With just a single frequency and one polarization, and capable of acquiring imagery at only one angle, SIR-A showed that in the dry Sahara Desert it could penetrate as deeply as 3 m. These early images from the dunes and drift sand of the eastern Sahara showed previously unknown buried valleys, geologic structures, and possible Stone Age occupation sites (McCauley et al. 1982). Radar responses from bedrock and gravel surfaces beneath wind-blown sand several meters thick delineated sand- and alluvium-filled valleys, some nearly as wide as the Nile Valley and perhaps as old as middle Tertiary. The now-vanished major river systems that carved these large valleys probably accomplished most of the erosional stripping of this extraordinarily flat, arid region. Stone Age artifacts associated with soils in the alluvium suggested areas that may have been sites of early human occupation. The presence of old drainage networks beneath the sand (Figure 6.4) provided a geologic explanation for the locations of many playas and present-day oases that have been centers of episodic human habitation.

The success of the mission paved the way for a follow-on, the SIR-B in 1984, which could collect data at more than one angle by mechanically tilting its antenna, and then the SIR-C (SIR-C/X-SAR) in April and October 1994. The synthetic aperture radar on board SIR-C was fully polarimetric, capable of both transmitting and collecting information at

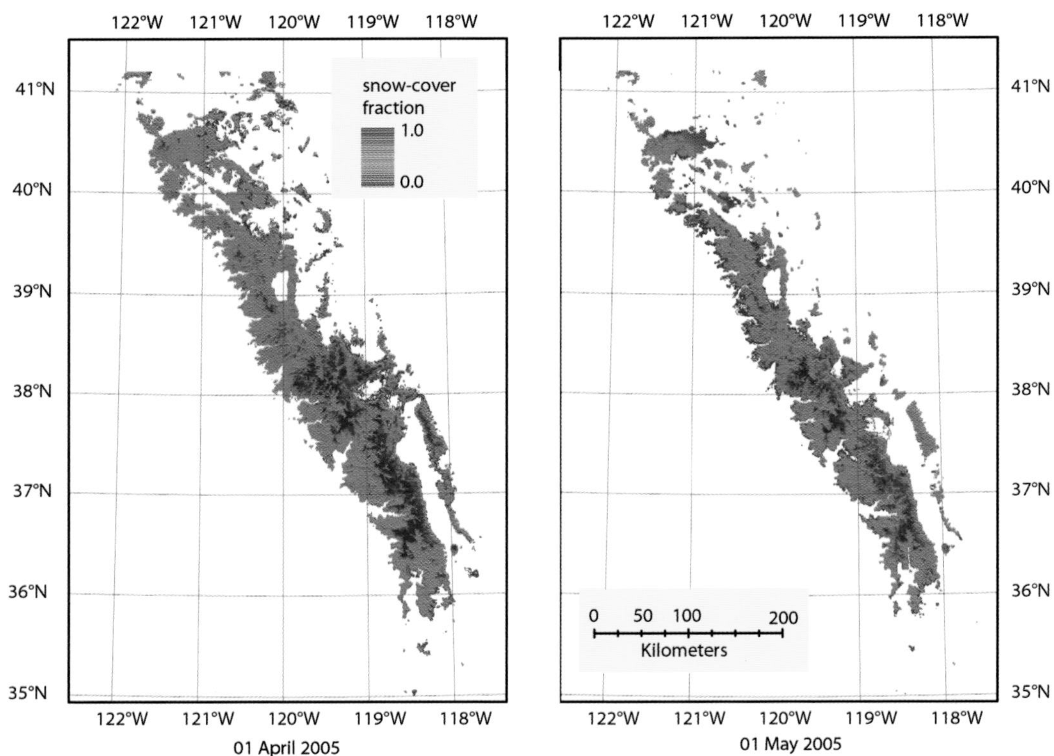

FIGURE 6.3 Fractional snow cover over the Sierra Nevada on April 1 (left) and May 1, 2005. Total snow-covered area is 23,100 km^2 in April and 14,900 km^2 in May. SOURCE: J. Dozier, University of California, Santa Barbara.

FIGURE 6.4 Image from Shuttle Imaging Radar-A (SIR-A) showing buried river channels in the Sahara Desert. SOURCE: *http://www.jpl.nasa.gov/history/hires/1981/SIR-A_image.jpg.*

vertical or horizontal polarizations. In addition, the antenna was electronically steerable and operated at three frequencies (1.4, 5.6, and 10 GHz). The two flights allowed investigation into the radar's response to seasonal changes. The multiparameter images were combined and enhanced to produce some of the most spectacular radar images ever seen.

ANALYSIS OF GROUNDWATER FROM GRAVITY DATA

A pair of satellites launched in 2002 makes up NASA's Gravity Recovery and Climate Experiment (GRACE). The main source of variation in Earth's gravity field is the movement of water between its three main reservoirs: the ocean, ice sheets, and groundwater. Unlike most satellite remote sensors, which measure electromagnetic radiation reflected or emitted from Earth's surface and atmosphere, GRACE measures the distance between its two spacecraft, which changes in response to variations in Earth's mass—and therefore gravity—on the surface below them. The GRACE measurement also senses mass change within the Earth—a capability demonstrated by the measurement of seasonal change in continental aquifers.

Wahr et al. (2004) use GRACE data to compare groundwater storage with a hydrologic model in the Mississippi and Amazon River basins and in the drainage flowing into the Bay of Bengal (Figure 6.5). When averaged over 1,000 km or more, the mass estimates inferred from the GRACE data clearly show annually varying changes in continental water storage, along with seasonal variability in the amount of water in the ocean. The amplitudes and phases of those signals are in general agreement with the hydrologic model. The inferred mass signals over the ocean are consistent with estimates of the water stored in the groundwater. Although the agreement degrades with decreasing averaging radius, the largest water storage signals are still clearly evident at averaging radii as short as 400 km. The globally averaged uncertainty in the amplitude of the annually varying mass signal recovered from these GRACE fields is 1.0 cm for a 1,000-km radius.

USE OF SATELLITE-DERIVED ELEVATION DATA IN HYDROLOGY

In February 2000, with the aid of a 60-m (200-ft) boom added to the SIR-C, the Shuttle Radar Topography Mission (SRTM) circled Earth for 10 days mapping 80 percent of the world's land area. The resulting high-resolution topographic map is the most accurate available and constitutes one of the major accomplishments of the nation's space program. SRTM (Farr et al. 2007) provides a worldwide topographic data set between 60° N and S latitudes with a consistent datum. Many areas otherwise lack topographic data, so these data enable spatial hydrologic modeling that would otherwise be impossible. Figure 6.6 shows an elevation and relief map of the whole African continent.

Since its launch, digital elevation models created from SRTM have been used in many applications, most notably tectonics, geomorphology, and hydrology. Because of their global consistency, SRTM data link continental hydrology with the oceans. In hydrologic investigations, the first information in characterizing a problem is often the topography of a drainage basin. From the elevations, slopes and aspects can be estimated, which are essential for calculations of solar and longwave radiation that can be used in spatially distributed energy balance models of snowmelt (Cline et al. 1998), photosynthesis, and evapotranspiration (Anderson et al. 2003). Analytical software packages use these slopes and aspects as input parameters to delineate drainage basin boundaries, to characterize basins for their distribution of slopes, and in routing water from precipitation or snowmelt (Tarboton 1997). An additional hydrologic application of SRTM data has been to measure water surface elevations directly (Alsdorf et al. 2007), which contributes to the improvement of flood forecasting.

Providing accurate flood forecasts from satellite observations is a high-priority mission with the potential to save lives and property (NRC 2007a). This important societal challenge

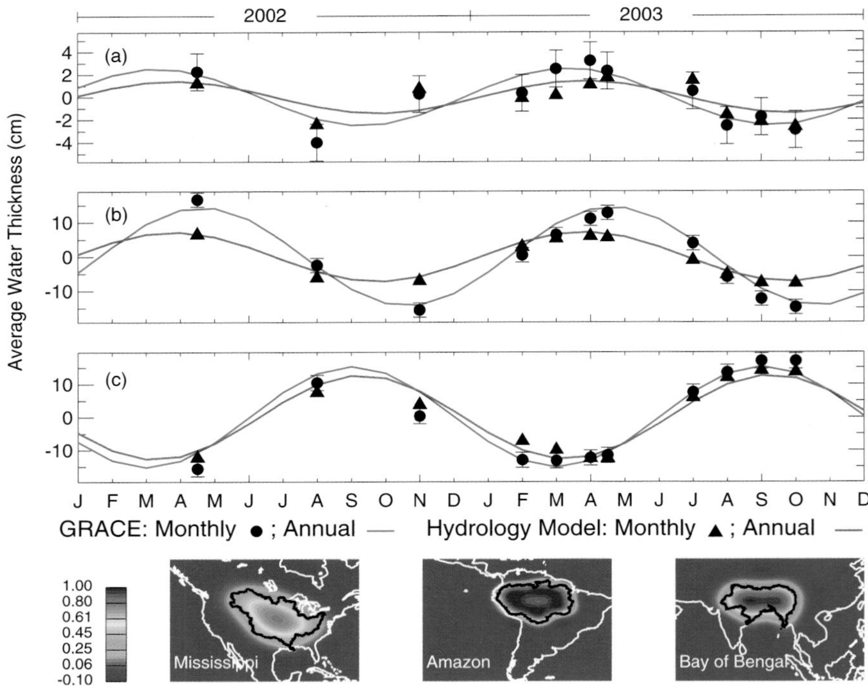

FIGURE 6.5 The mass variability within (a) the Mississippi River basin, (b) the Amazon River basin, and (c) a drainage system flowing into the Bay of Bengal, as inferred from the GRACE measurements (dots). Also shown are results inferred from a hydrologic model, as well as the best-fitting annual signal for both the GRACE values and the model predictions. Bottom panels show the optimal averaging kernels used to recover this mass variability. SOURCE: Wahr et al. (2004). Reprinted with permission from the American Geophysical Union, copyright 2004.

cannot be answered adequately with the current global in situ networks designed to observe river discharge. Knowledge of soil moisture (and snow water storage, where relevant), surface water area, the elevation and slope of the water surface, and accurate hydrologic models is required to meet this challenge. Although attempts to estimate soil moisture from the AMSR-E sensor have been made, they are only at the early experimental stage. Nevertheless, results have shown promises and the proposed Soil Moisture Active-Passive mission is central to making progress toward reliable flood hazard assessments (NRC 2007a). Despite the many accomplishments highlighted in this chapter, important challenges remain such as the GPM, soil moisture estimates, surface water and ocean topography (to improve estimates of water stored in lakes, reservoirs, wetlands, and rivers), and improved estimates of snowpacks (NRC 2007a).

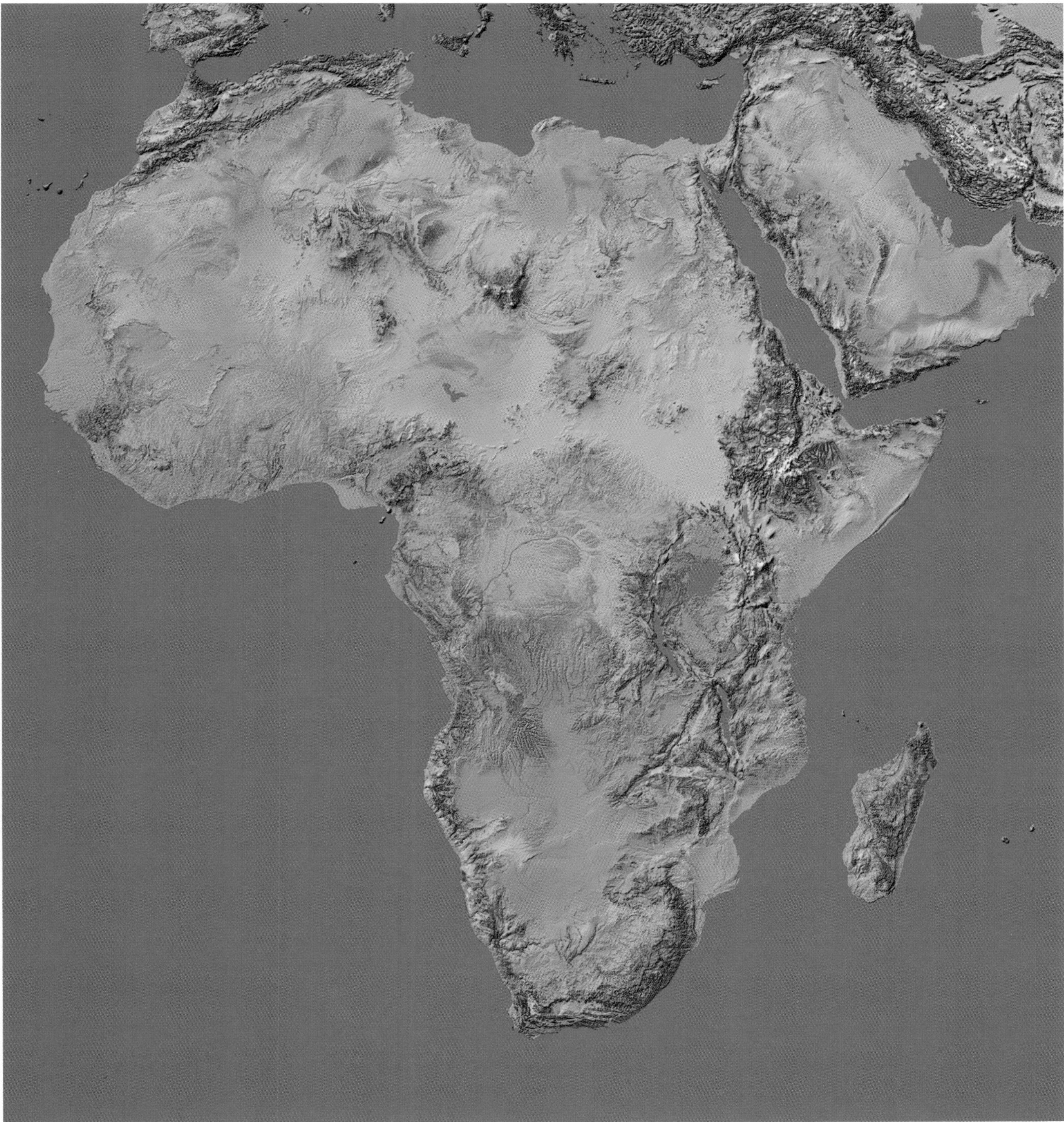

FIGURE 6.6 Elevation and relief map of Africa from the Shuttle Radar Topography Mission. Color coding is directly related to topographic height, with brown and yellow at the lower elevations, rising through green, to white at the highest elevations. Blue areas on the map represent water within the mapped tiles, each of which includes shorelines or islands. SOURCE: NASA Jet Propulsion Laboratory.

7

Cryosphere

No other single technological development has revolutionized cryosphere research as much as satellite observations. Most of Earth's frozen regions are remote and access by land or seas involves often great risks; therefore, conducting in situ observations is logistically difficult and expensive. The synoptic view from satellites increases the data coverage by multiple orders of magnitude, and access is no longer restricted by seasons.

Understanding changes to ice sheets, sea ice, ice caps, and glaciers is important for understanding global climate change and predicting its effects. In particular, "shrinking ice sheets" and their contribution to sea-level rise were identified as the third most significant "Breakthrough of the Year" for 2006 according to *Science* magazine[1]:

> Glaciologists nailed down an unsettling observation this year: The world's two great ice sheets—covering Greenland and Antarctica—are indeed losing ice to the oceans, and losing it at an accelerating pace. Researchers don't understand why the massive ice sheets are proving so sensitive to an as-yet-modest warming of air and ocean water. The future of the ice sheets is still rife with uncertainty, but if the unexpectedly rapid shrinkage continues, low-lying coasts around the world—including New Orleans, South Florida, and much of Bangladesh—could face inundation within a couple of centuries rather than millennia.
>
> —*Science* (2006)

This breakthrough is one of the examples of major accomplishments in cryosphere science presented in this chapter. Other examples include the change in seasonal snow cover (see Chapter 6), detection of earlier spring thaw and associated lengthening of the growing season (Chapter 9, Box 9.4), the new perspective of the dynamic ice streams in Antarctica, the decrease in sea ice in the Arctic, and the change in glacier extent.

NONUNIFORM AND DYNAMIC ICE STREAMS IN ANTARCTICA

Field exploration of the Antarctic ice sheet is time consuming, logistically intensive, costly, and sometimes dangerous. Prior to satellite observations, spatial coverage was very sparse: information on the Antarctic ice sheet was acquired slowly over the years by numerous surface traverses (Figure 7.1).

In 1997, Radarsat data were used to create the first complete radar-based map of Antarctica (Figure 7.1). Analyses of radar images from various sensors through the years have enabled detailed measurements of surface velocity, and in turn these measurements have enabled calculation of strain rates and basal shear at the bed. For the first time, satellite data revealed the extent of the ice stream network, leading to the discovery of new ice streams and the ice stream tributaries (Joughin et al. 1999). For example, satellite-based measurements of surface velocity within Antarctic ice streams reveal a complex pattern of flow not apparent from previous measurements (Bindschadler et al. 1996). Furthermore, satellite observations led to the discovery that ice streams move at variable speeds, resulting in a more dynamic picture than the previously held view that ice sheets move at a constant velocity (Figure 7.2; Bindschadler and Vonberger 1998). Satellites provide improved data collection methods to increase data density and to improve velocity estimates substantially.

[1] The first-ranking breakthrough of the year was the proof of the Poincaré conjecture, a long-standing problem in mathematics. The second-ranking breakthrough was in the area of paleogenomics: the sequencing of Neanderthal DNA proves that Neanderthal evolution diverged from modern humans at least 450,000 years ago.

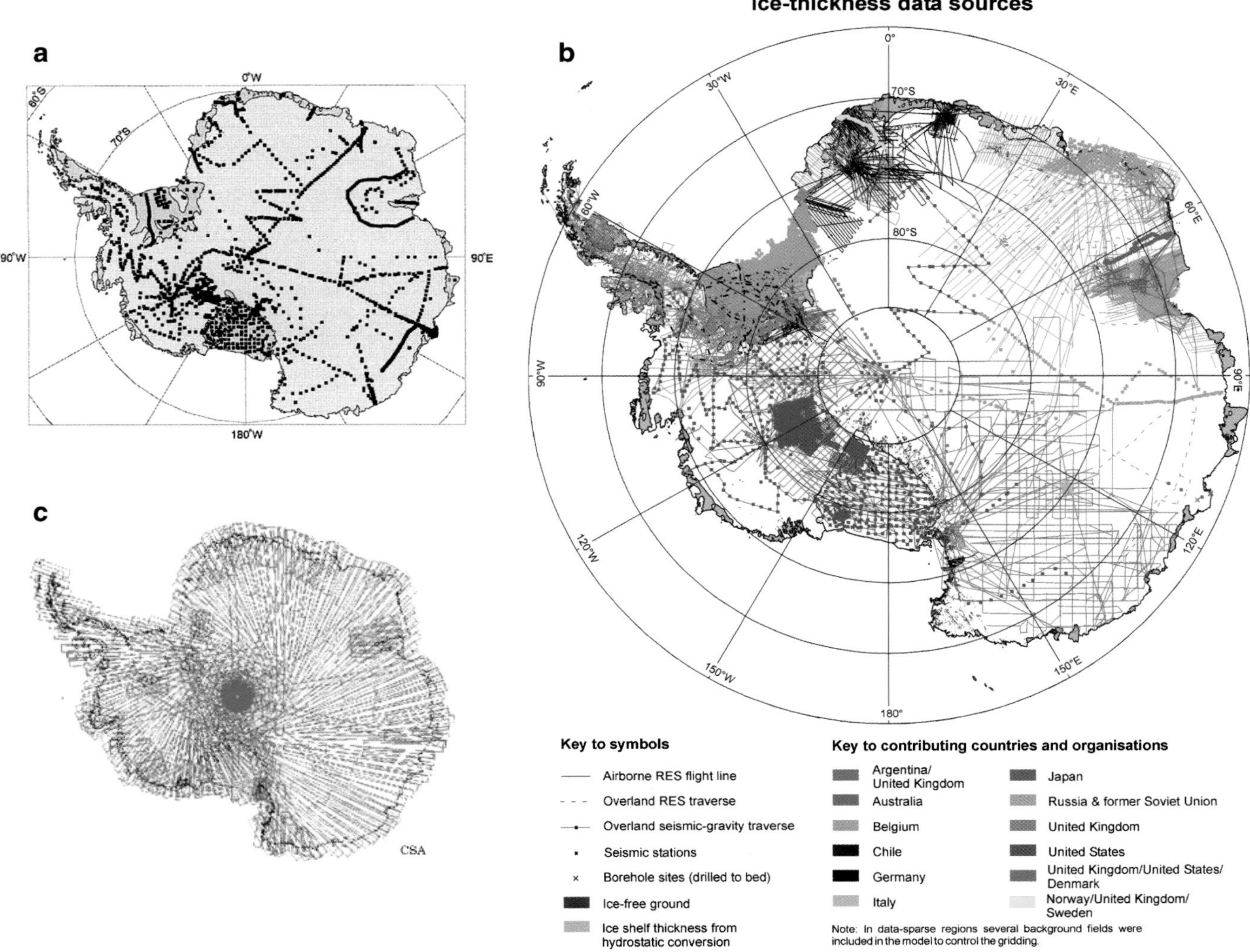

FIGURE 7.1 Spatial coverage of data from Antarctica. (a) Surface transverses since the 1957-1958 International Geophysical Year. SOURCE: National Snow and Ice Data Center, University of Colorado. (b) Airborne surveys, over snow radio-echo sounding (RES), seismic surveys, gravimetric surveys, and ice-core missions since the 1957-1958 International Geophysical Year. SOURCE: BEDMAP consortium. (c) Satellite coverage. SOURCE: John Crawford, Canadian Space Agency, National Aeronautics and Space Administration, Jet Propulstion Laboratory.

ACCELERATING ICE SHEET FLOW IN ANTARCTICA AND GREENLAND

One of the central questions in climate change and cryosphere research is how the warming climate will affect the ice sheets because the amount of continental ice and melt water entering the ocean strongly contributes to the change in sea level. Glaciologists and climatologists have long been debating whether a warming climate would decrease ice mass. However, early research focused on how increased melting would be offset by increased precipitation. The ice mass balance and thus its contribution to sea-level rise was originally thought to be determined by the difference between melting and precipitation.

Satellite observations have revolutionized this thinking by allowing scientists to monitor precise ice sheet elevation, velocity, and overall mass. Satellite images revealed that in fact the overall mass is declining (Luthcke et al. 2006). In addition to observing great variability in the ice stream velocity over time and space, satellite images revealed that the overall velocities of the ice streams in Antarctica and Greenland have increased during the past decade, resulting in more ice flow into the ocean (Bindschadler and Vonberger 1998, Joughin et al. 2001).

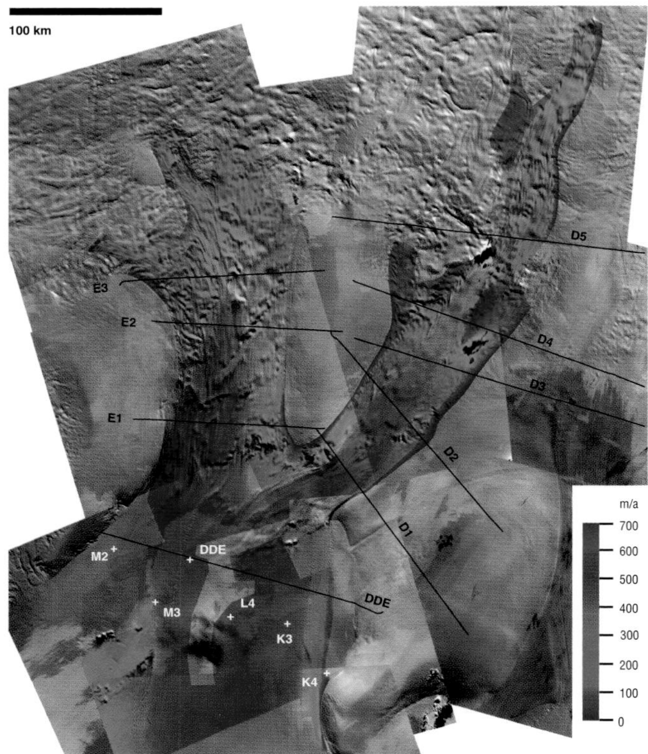

FIGURE 7.2 Velocity variations in Antarctica ice streams. SOURCE: Binschadler et al. (1996). Reprinted from the Annals of Glaciology with permission of the International Glaciological Society, copyright 1996.

The discoveries of accelerating ice loss from Antarctica and Greenland and the importance of ice sheet dynamics in their mass balances rest on measurements by a suite of satellite and airborne sensors using novel techniques (Bindschadler et al. 1996, Chen et al. 2006a, Kerr 2006, Luthcke et al. 2006, Rignot and Kanagaratnam 2006). These discoveries are possible because of decades of optical and radar images, laser and radar altimeters, and more recently the National Aeronautics and Space Administration's (NASA) Gravity Recovery and Climate Experiment (GRACE) mission, which measures ice mass directly through its gravitational pull. In addition, airborne laser altimeter data show thinning of ice near the coastline, radar data show faster flow, Landsat data show retreat of the grounding line,[2] and the Moderate Resolution Imaging Spectroradiometer (MODIS) data show calving of large icebergs. Warming ocean waters seem to have increased calving of the ice shelves, thereby allowing the ice sheet's outlet glaciers to flow more quickly (Box 7.1). Glaciers in Greenland have also increased in velocity, perhaps from increased basal lubrication by meltwater penetrating from the surface. These new discoveries indicate that ice stream dynamics (the balance between the forcing, such as ice thickness and surface slope, and the resistance, such as internal stiffness) are the primary drivers of rapid sea-level change instead of the balance between melting and precipitation.

The ability to estimate the overall mass of ice sheets is a remarkable accomplishment of satellite observations. Numerous techniques, including radar images, measurements of surface elevation from laser altimeters, and GRACE's gravity data, now show that both Greenland and Antarctica have been losing ice over the past 5 to 10 years. From 2003 to 2005, Greenland lost more than 155 gigatons[3] per year at lower elevations and gained about 54 gigatons per year at higher elevations, with most of the losses occurring during summer (Chen et al. 2006b, Luthcke et al. 2006, Rignot and Kanagaratnam 2006, Wahr et al. 2006). In Antarctica the gravity data show mass losses of 70-200 km^3 per year (60-160 gigatons per year). Most of the loss is from West Antarctica, with East Antarctica in approximate balance (Figure 7.3).

DECLINING ARCTIC SUMMER SEA ICE

> Just as miners once had canaries to warn of rising concentrations of noxious gases, researchers working on climate change rely on arctic sea ice as an early warning system.
>
> —*Arctic Climate Impacts Assessment* (2004)

For many reasons, observing trends in sea ice reliably has been possible only with the advent of satellite observations. Navigating the remote and frozen seas off Antarctica or in the Arctic to obtain in situ measurements of sea ice extent is treacherous, and sea ice extent is highly variable in time and space due to wind advection and localized melting. Before satellite observations became available, spatial coverage of sea ice was monitored by tracking the location of the ice edge from ships. Because the ice edge is moving with winds and ocean currents, it is not a robust indicator of basin-scale sea ice extent. Thus, accurate and quantitative interannual comparisons of the basin-scale ice coverage became only possible with the availability of the synoptic view from satellites.

Sea ice has been monitored continuously with passive microwave sensors (Electrically Scanning Microwave Radiometer [ESMR], Scanning Multichannel Microwave Radiometer [SMMR], Special Sensor Microwave/Imager [SSM/I], and Advanced Microwave Scanning Radiometer-Earth Observing System [AMSR-E]) since 1979. Not limited by weather conditions or light levels, they are particularly

[2]The location along the coast where ice is no longer supported by the ground and where it begins to float.

[3]1 gigaton = 1 billion metric tons.

BOX 7.1
Ice Shelf Collapse

The observation of the collapse of the Larsen B Ice Shelf was astonishing in the sheer dimension and abruptness of change observed via satellite, and it alerted the Earth science community due to its potential implications for sea-level rise (Figure 7.3). The dynamics contributing to the collapse were documented by various satellites: the thinning of the ice shelf toward the coast by satellite altimetry, the accelerated flow by the interferometric synthetic aperture radar (InSAR), the retreat of the grounding line by Landsat, and the calving of the icebergs by MODIS.

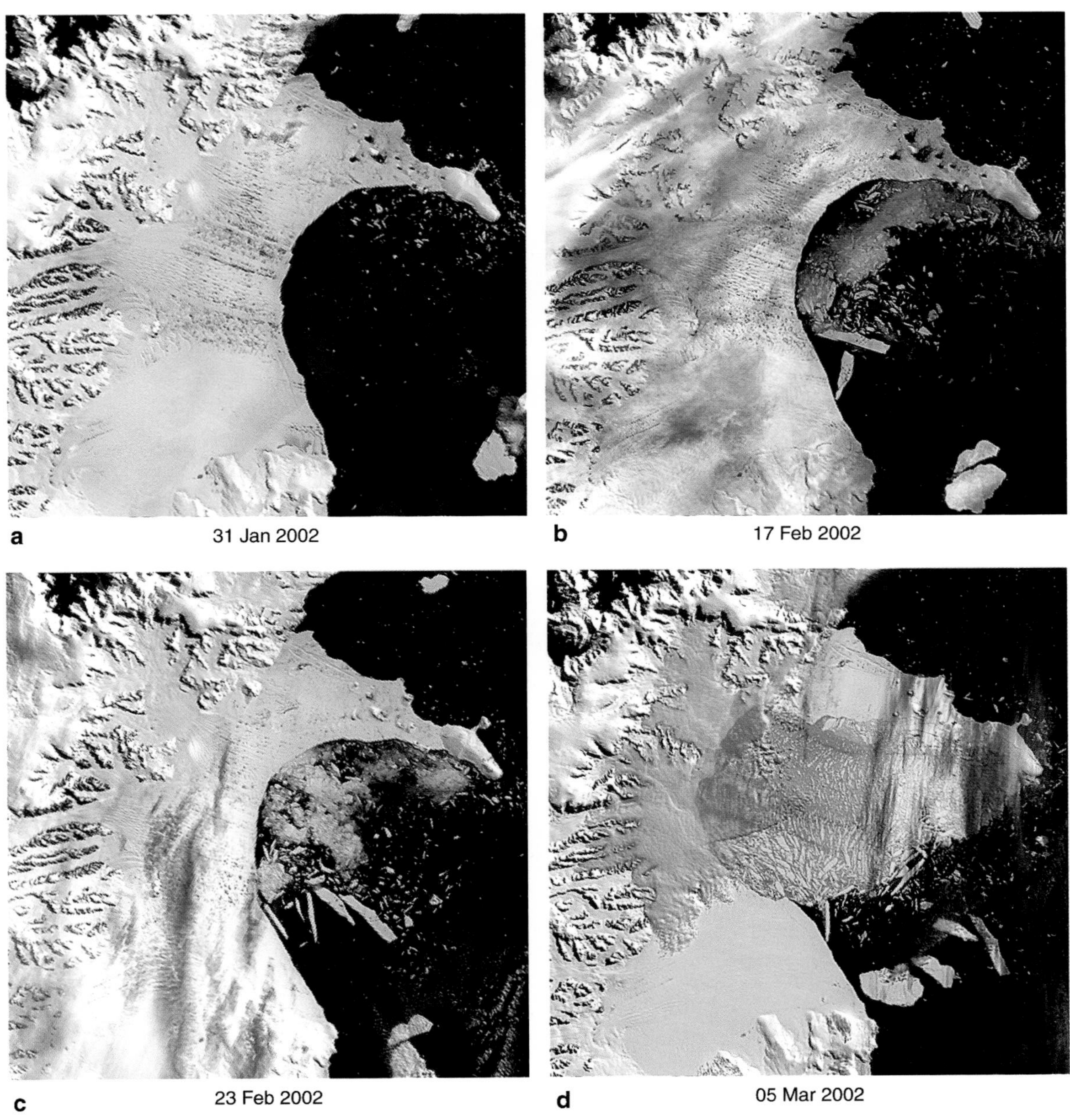

FIGURE 7.3 Collapse of the Larson B Ice Shelf in western Antarctica, January-March 2002. Two thousand square kilometers of the Larsen Ice Shelf disintegrated in just 2 days. SOURCE: National Snow and Ice Data Center, University of Colorado.

well suited for monitoring sea ice because of the strong contrast in microwave emission between open and ice-covered ocean. The long-term 35-year data set from the passive microwave sensors has enabled us to produce trend analyses beyond the strong interannual variability of sea ice. Recent estimates indicate that Arctic sea ice extent decreased by approximately 7.4 percent from 1978 through 2003, while multiyear ice area has decreased by approximately 7.0-11.0 percent per decade (Comiso 2002, Johannessen et al. 2004; Figure 7.4). The past several years have been nothing short of extraordinary (NRC 2007c). Since 2000, record summer ice minima have been observed during 4 out of the past 6 years in the Arctic (Stroeve et al. 2005). Moreover, most recent indications are that winter ice extent is now also starting to retreat at a faster rate, possibly as a result of the oceanic warming associated with a thinner, less extensive ice cover. These observations of shrinking Arctic sea ice are consistent with climate model predictions of enhanced high-latitude warming, which in turn are driven in significant part by ice-albedo feedback[4] (Holland and Bitz 2003). In contrast to the Arctic, no clear trend in the extent of Antarctic sea ice coverage has been detected.

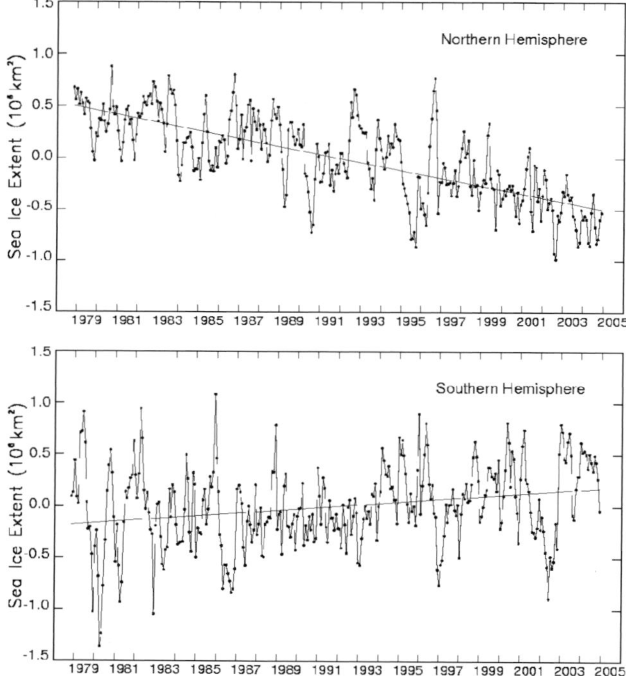

FIGURE 7.4 Deviations in monthly sea ice extent for the northern and southern hemispheres from November 1978 through December 2004, derived from satellite passive-microwave observations. The Arctic sea-ice decreases are statistically significant, with a trendline slope of $-38,200 \pm 2,000$ km^2/year, and have contributed to much concern about the warming Arctic climate and the potential effects on the Arctic ecosystem. The Antarctic sea ice increases are also statistically significant, although at a much lower rate of $+13,600 \pm 2,900$ km^2/year. The northern hemisphere plot is extended from Parkinson et al. (1999), and the Southern Hemisphere plot is extended from Zwally et al. (2002). SOURCE: Courtesy of Claire Parkinson and Donald Cavalieri, NASA Goddard Space Flight Center, as updated from Parkinson et al. (1999) and Zwally et al. (2002).

Over the past few years, there have been a growing number of reports forecasting sea ice conditions, and these reports are based entirely or mostly on data from satellites. For example, the Arctic Climate Impact Assessment (ACIA 2005) concluded that continued reductions in Arctic sea ice might soon lead to a seasonally ice-free Arctic and increased maritime traffic because shipping routes through the Arctic Ocean are much shorter than routes through the Panama or Suez Canals. However, there is some evidence that a reduction in the ice cover will be accompanied by greater interannual variability, at least in certain regions (Atkinson et al. 2006); the potential combination of increased maritime traffic, high interannual variability in the ice cover, and regional variations will require improved regional sea ice forecasts for maritime operators.

GLACIER EXTENT AND POSITION OF EQUILIBRIUM LINE

The study of glacier regimes worldwide reveals widespread wastage since the late 1970s, with a marked acceleration in the late 1980s. Remote sensing is used to document changes in glacier extent (the size of the glacier) and the position of the equilibrium line (the elevation on the glacier where winter accumulation is balanced by summer melt; König et al. 2001). Since 1972, satellites have provided optical imagery of glacier extent. The synthetic aperture radar (SAR) is used to study zones of glacial snow accumulation and ice melt to determine climate forcing, and laser altimetry is used as well to measure change in glacier elevation. For example, a study in the Ak-shirak Range of the central Tien Shan plateau used aerial photographs in the 1970s, along with the Advanced Spaceborne Thermal Emission and Reflection Radiometer (ASTER) imagery from 2001, to document a reduction in glacier area of 20 percent between 1977 and 2001 (Figure 7.5; Khromova et al. 2003).

Because glaciers respond to past and current climatic changes, a complete global glacier inventory is being developed to keep track of the current extent as well as the rates of change of the world's glaciers. Coordinated by the National Snow and Ice Data Center, the Global Land Ice Measurements from Space project is using data from ASTER and the Landsat Enhanced Thematic Mapper to inventory about 160,000 glaciers worldwide. This effort will likely

[4]Ice-albedo feedback is a positive feedback loop whereby melting sea ice exposes more seawater (of lower albedo, or less reflective), which in turn absorbs heat and causes more sea ice to melt.

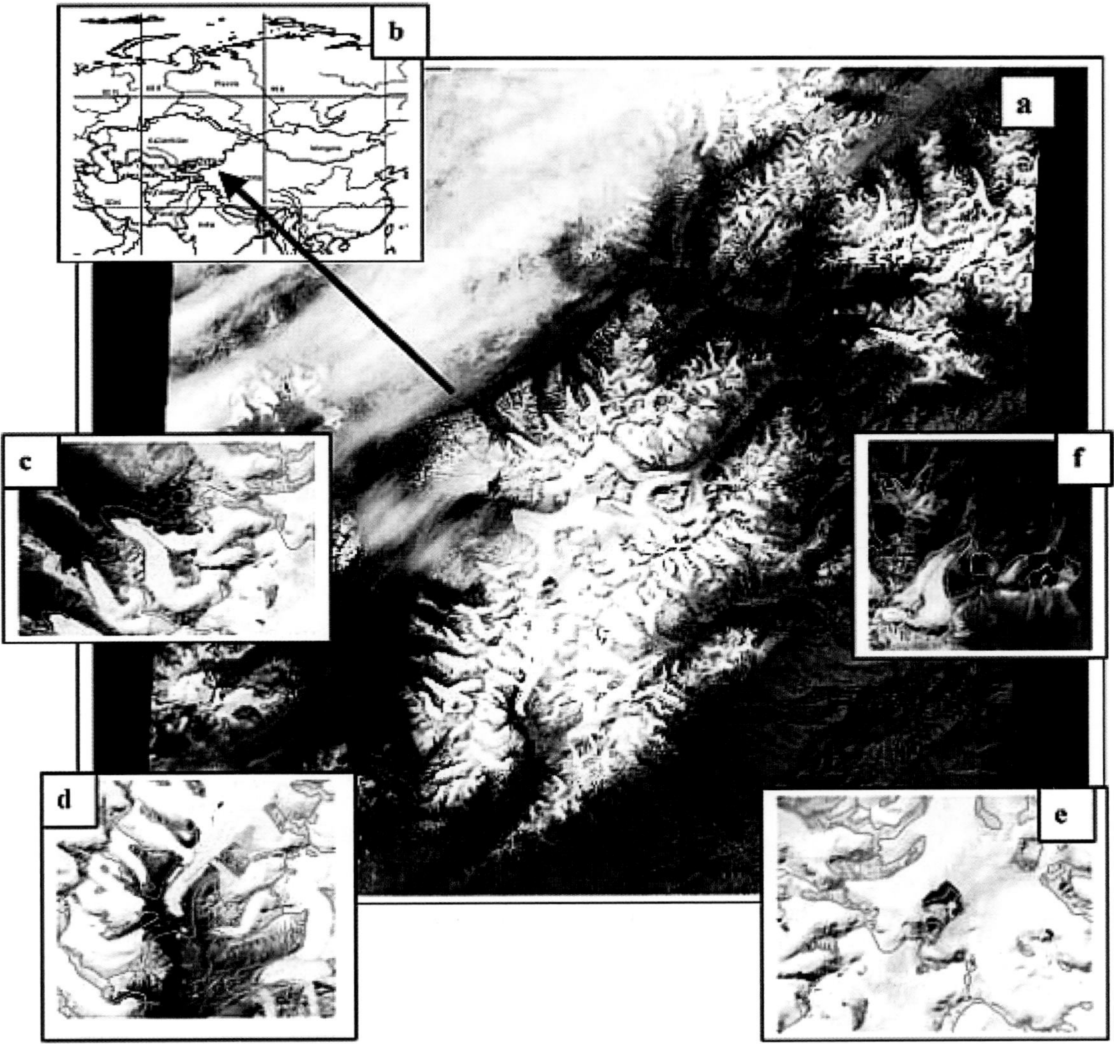

FIGURE 7.5 Location and changes of the Ak-shirak glacier system, 1943-2001. (a) ASTER image for September 14, 2001. (b) Location map. In the other insets the green lines indicate glacier outlines in 1943: (c) decrease in glacier size through climate change and direct anthropogenic impact, (d) decrease in size of a surging glacier and appearance of new glaciers, (e) increase in area of outcrops and in the perimeters of water divides between glaciers, and (f) disappearance of former small glaciers. SOURCE: Khromova et al. (2003). Reprinted with permission from the American Geophysical Union, copyright 2003.

result in major scientific advances in the near future with important ramifications for climate research. As for the other examples of accomplishments listed in this chapter, these measurements and the resulting trend analyses are important indicators of climate change and exemplify the value and importance of long-term data sets for understanding the complex climate system.

8

Ocean Dynamics

Due to the remoteness of the vast open oceans, satellites provided the first truly global ocean-observing system. Presatellite observing platforms included ships, moorings, drifters, and other tools, none of which could provide ocean basin-scale coverage at the temporal and spatial scales required to resolve the dynamic nature of the ocean that has been revealed since. In fact, even a well-known and studied current such as the Gulf Stream was not fully characterized until satellite observations were available (Box 8.1, Figures 8.1 and 8.2). Satellite data from scatterometers, altimeters, infrared radiometers, and various ocean color sensors opened up a new window for observing and quantifying how and why water moves around in the ocean (ocean dynamics) and how energy is exchanged between the ocean and atmosphere (air-sea interaction).

As illustrated in more detail below, sea surface temperature (SST) measurements not only revealed important information about ocean circulations (e.g., the Gulf Stream) but also advanced climate research by providing detailed information on the heat input into the ocean. Ocean color combined with SST observations led to new discoveries about the physical-biological coupling in the ocean, with important implications for the ocean's role in the carbon cycle (see also Chapter 9). Observations from altimeters have resulted in a slow revolution, as the accuracy of the sensors steadily increased, taking about a decade for their contributions to be broadly recognized. Altimetry embedded within other modern ocean measurements and models yielded a virtually complete description of first-order physical processes in the ocean. The measurements provided real surprises to physical oceanographers, including detection of the internal tide in the open ocean, of a highly variable surface ocean full of eddies, and of global sea-level trends at an accuracy of millimeters per year. Combining satellite data with in situ observations and models converted physical oceanography into a global science with actual predictive skill.

THE OCEAN'S ROLE IN CLIMATE CHANGE

Monitoring SST by the Advanced Very High Resolution Radiometer (AVHRR) is the marine remote sensing technique with the broadest impact on oceanography (Robinson 1985). SST is the earliest Earth-orbiting satellite measurement for oceanography and began with the launch of the Television Infrared Observation Satellite (TIROS-N) in 1978. SST measurements provide the longest continuous record of any oceanic property from space (Table 8.1). This long-term data set has been calibrated and validated by surface observations from drifters, buoys, and ships and is used for a broad range of oceanographic research questions, including studies of regional climate variability, most notably El Niño-Southern Oscillation (ENSO; see also Chapter 12), climate change, and ocean currents.

SST is one of the most important indicators of global climate change and a vital parameter for climate modeling (Hurrell and Trenberth 1999). Because of the large heat content of the ocean, more than 80 percent of the total heating of the Earth system is stored in the ocean, and ocean currents redistribute this heat across the globe. Consequently, "if [scientists] wish to understand and explain [global] warming, the oceans are clearly the place to look" (Barnett et al. 2005b). In addition, SST is central in coupling the ocean with the atmosphere and is a controlling factor in the heat and vapor exchange between the two (Johannessen et al. 2001). Trend analysis of SST provided evidence for global warming and the important climate-atmosphere feedback in the tropics that is also responsible for ENSO events (Cane et al. 1997). These SST observations, combined with in situ vertical temperature measurements of the ocean to a depth of 3,000 m provided evidence to detect anthropogenic global warming in the ocean (Barnett et al. 2001, 2005b).

Understanding the increase in SST and anthropogenic heat input to the surface ocean also has important

BOX 8.1
Gulf Stream Path

Describing the path of the Gulf Stream and other major ocean currents was an early challenge to physical oceanographers who based their interpretations on very sparse data collected from oceanographic ships. For example, using data from a multiship survey, Fuglister and Worthington (1951) proposed the Multiple Current Hypothesis, which suggested that an instantaneous chart of the Gulf Stream would show a number of disconnected filaments of current that change in time (Figure 8.1). Furthermore, they concluded that three Gulf Stream configurations were possible: a single filament (Figure 8.1a), a branching current with two filaments (Figure 8.1b), or a number of irregular disconnected filaments (Figure 8.1c). In subsequent years, ship data could not distinguish between these three and other interpretations. However, in the mid-1970s the synoptic view provided by satellite thermal infrared imagery showed that the Gulf Stream was a single filament, albeit following a tortuous and time-changing path (Figure 8.2).

Over many years synoptic views of the Gulf Stream were obtained via satellite radiometers. These results showed considerable interannual variability in the path of the stream based on the position of the "North Wall"—the boundary at which strong temperature gradients (fronts) between warm Gulf Stream waters and colder waters of the Northwest Atlantic demarcate the northernmost extent of the stream (Lee and Cornillon 1995). These interannual motions were subsequently shown to be important to fisheries (Olson 2001) and to the productivity of the Slope Sea (Schollaert et al. 2004).

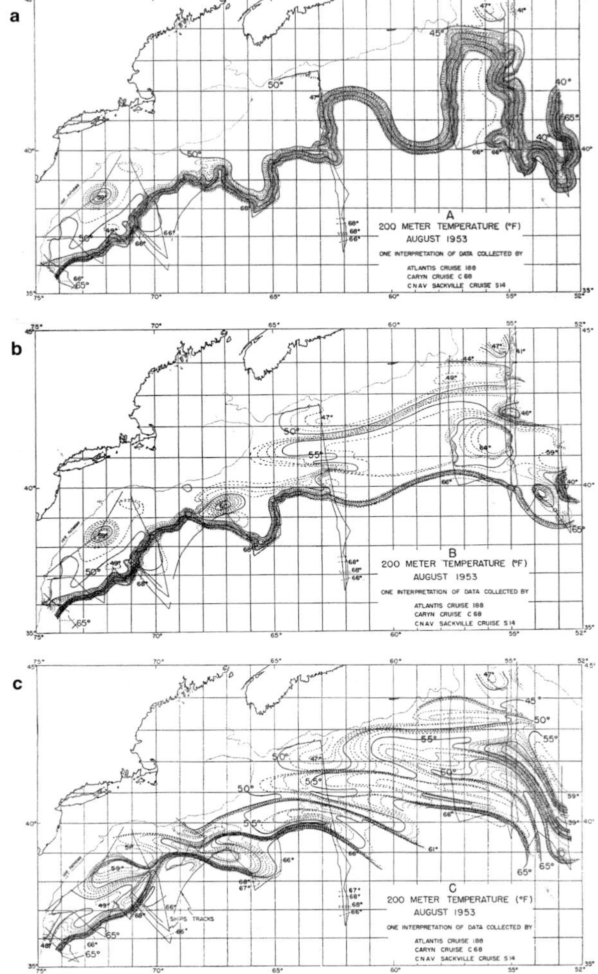

FIGURE 8.1 Fuglister's multiple current hypothesis. SOURCE: Stommel (1965). Reprinted with permission from the University of California Press, copyright 1965.

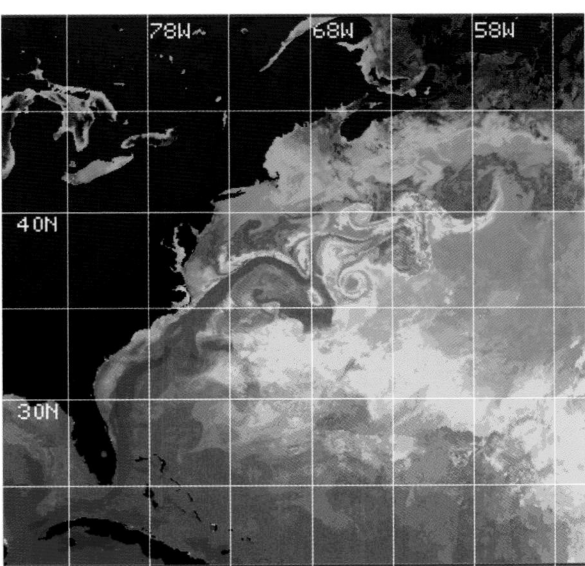

FIGURE 8.2 SST image showing the Gulf Stream in the Atlantic Ocean. SOURCE: Provided by Otis Brown and Bob Evans.

TABLE 8.1 TIROS-NOAA Satellites Carrying AVHRR Sensors Monitoring SST

Satellite	Dates of Operation
TIROS-N	Oct. 1978-Jan. 1980
NOAA-6	June 1979-March 1983
NOAA-7	Aug. 1981-Feb. 1985
NOAA-8	May 1983-Oct. 1985
NOAA-9	Feb. 1985-Nov. 1988
NOAA-10	Nov. 1986-Sept. 1991
NOAA-11	Nov. 1988-April 1995
NOAA-12	Sept. 1991-present
NOAA-14	Dec. 1994-present
NOAA-15	May 1998-present
NOAA-16	Sept. 2000-present

ramifications for quantifying and predicting sea-level rise (Cabanes et al. 2001). Recent work suggests that thermal expansion of the surface ocean (upper 500 m) can fully explain the sea-level rise of 3.2 (± 0.2) mm per year observed by the satellite TOPEX/Poseidon (Cabanes et al. 2001).

PREVALENCE OF DYNAMIC FEATURES

The ability to observe the ocean surface from space has profoundly altered the way the ocean is viewed. The Coastal Zone Color Scanner (CZCS), launched aboard the Nimbus 7 satellite in 1978, provided the first satellite observations used to quantify the chlorophyll concentration in the upper ocean (see Chapter 9, Box 9.3). The first images of surface chlorophyll distributions were truly astonishing, revealing a high degree of spatial variability never fully appreciated before satellites (Figure 8.3). The availability of global maps of chlorophyll, an estimate for marine plant biomass, has opened new avenues of research and changed the conduct of biological oceanography in many ways.

Mesoscale features such as vortices and jets, as well as tidal fronts and river plumes, had been seen previously in aerial photographs and thermal imagery from the TIROS satellites, but ocean color images revealed entirely new features. An example of this is the vast extent of the Amazon River plume stretching many thousands of kilometers across the Atlantic (Figure 8.4, Muller-Karger et al. 1988). The plume's temperature does not provide sufficient contrast

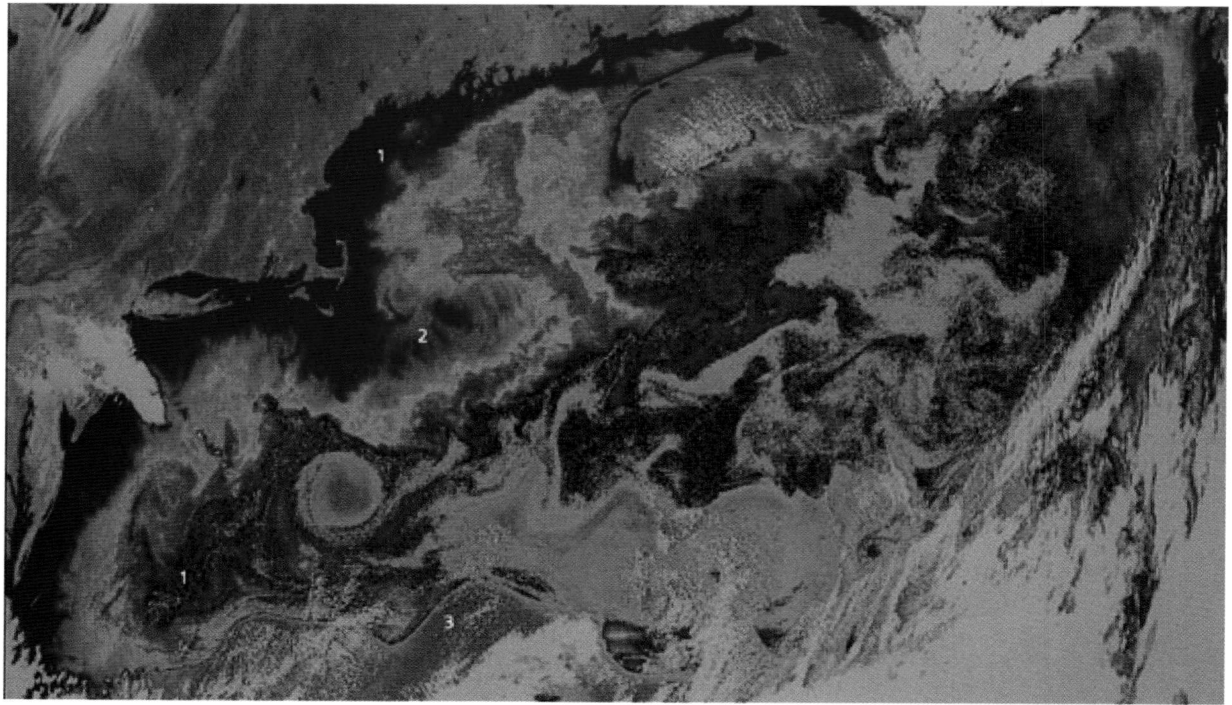

FIGURE 8.3 CZCS image of phytoplankton pigments in the North Atlantic Ocean. CZCS was flown on the Nimbus 7 satellite launched in 1978. CZCS was the first multispectral imager designed specifically for satellite observations of ocean color variations. One of the primary determinants of ocean color is the concentration of chlorophyll pigments in the water. High concentrations of chlorophyll (red and brown areas in the image) are seen along the continental shelf (1) and above Georges Bank (2) where the biological productivity is high. Intermediate concentrations of chlorophyll pigments are shown in green, and the lowest levels are blue. Notice that the Gulf Stream (3) and the warm core eddy to the north (blue circle) have very low concentrations, reflecting the fact that the stream and the Sargasso Sea to the south are relatively nutrient poor. SOURCE: NASA.

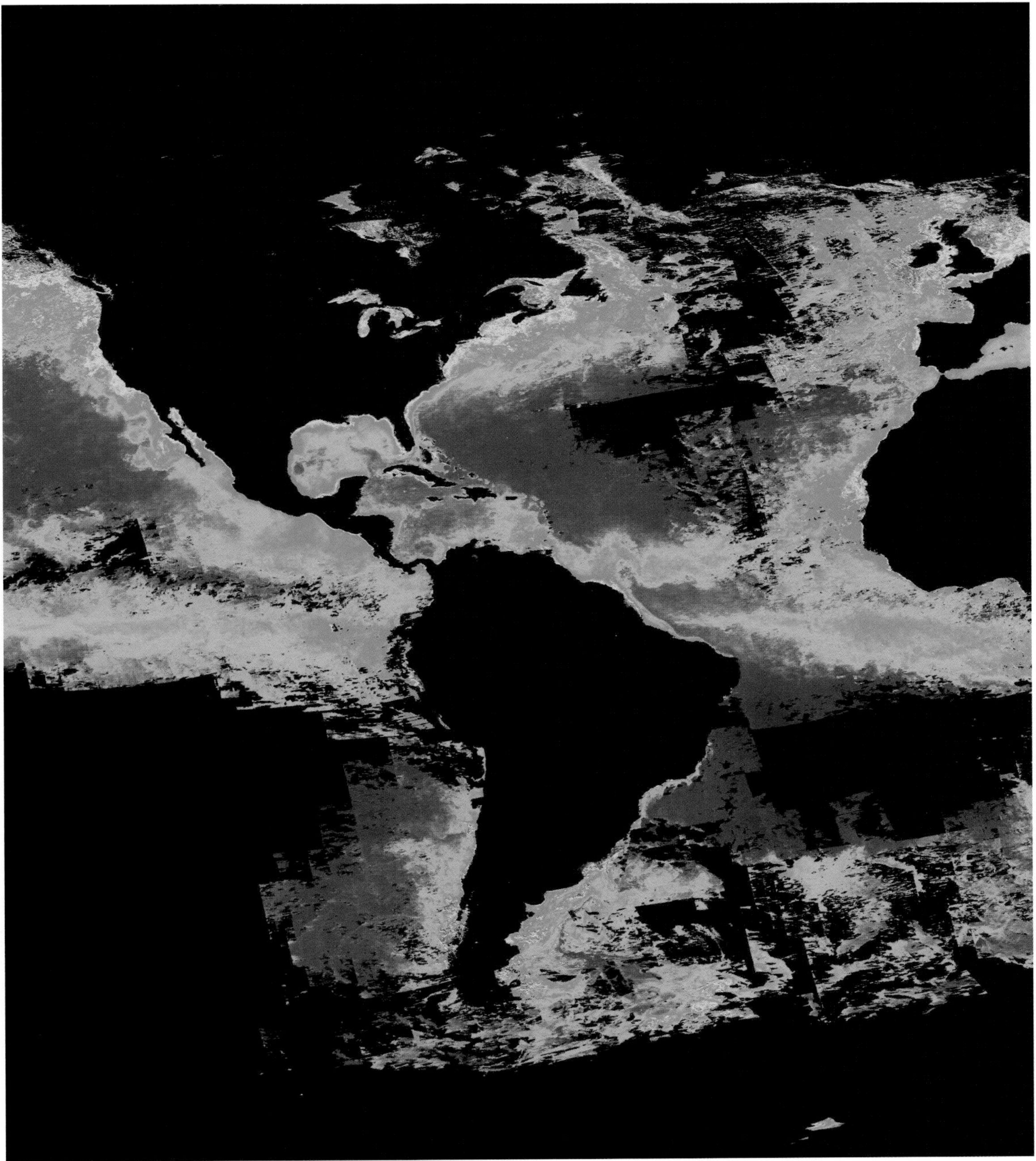

FIGURE 8.4 The vast extent of the Amazon River plume stretching thousands of kilometers into the Atlantic Ocean is an example of a new discovery that resulted from the first ocean color observations from space (Muller-Karger et al. 1988). The Amazon River plume is the green band extending across the Atlantic in this seasonally averaged CZCS pigment image for the months of September to November 1979. Bands of high pigment also mark the nutrient-rich upwelling along the equator in the Pacific and Atlantic, and the high latitudes and coastal regions are also seen as productive. Black areas over the ocean are missing data because CZCS operated only intermittently. SOURCE: SeaWiFS Project, NASA Goddard Space Flight Center, and GeoEye. Provided by the SeaWiFS Project, NASA/Goddard Space Flight Center and GeoEye.

with the tropical Atlantic to be visible in thermal imagery, whereas its color makes it clearly visible. River discharge measurements from gauging stations have been shown to correlate with the temporal variability of plumes in satellite images of the Gulf of Mexico (Salisbury et al. 2001, 2004), thus offering a method for studying the influences of rivers on the coastal ocean.

As a result of satellite images, Earth scientists have gained a physical perspective and appreciation of the relationship of the ocean to land masses. Seaward-flowing jets and filaments associated with major fronts along the continental shelf off California and the Pacific Northwest were a focus of the Coastal Transition Zone Program during the CZCS era (Brink and Cowles 1991). These narrow filaments of productive water extending hundreds of kilometers seaward from the continental margin are now recognized as important pathways for the transport of materials from the continental shelves to the deep ocean (Strub et al. 1991). Other researchers used CZCS to look at the Columbia River plume (Fiedler and Laurs 1990) and to relate tuna catch to fronts and features seen in satellite images (Laurs et al. 1984).

UNDERSTANDING OCEAN TIDES: NEW SOLUTIONS TO AN OLD SCIENTIFIC QUESTION

Ocean tides have fascinated scientists since the early Greeks and were first explained by Newton to be caused by the gravitational attraction of the Moon and the Sun. A century later Newton's theory was replaced by the dynamic response concept described by Laplace's (1776) tidal equation. Because Laplace's tidal equation strongly depends on the shape and bathymetry of the ocean basin and because oceans have clusters of natural resonances in the same frequency bands as the gravitational forcing function (Platzman 1981), analytical solutions to Laplace's tidal equation cannot be found. Therefore, predicting ocean tides to some level of accuracy was made possible only by Darwin's (1886) empirical method. The behavior of ocean tides, particularly in the open ocean, remained elusive until the advent of satellite altimetry (Le Provost 2001).

For the first time, satellite altimetry observations allowed synoptic measurements of ocean tides in the global open ocean. Although the first altimetry data were obtained from Geodynamics Experimental Ocean Satellite 3 (Geos3) in 1973, it was not until Seasat in 1978 that it became evident that a tidal signal could be retrieved from satellite altimetry (Le Provost 1983, Cartwright and Alcock 1983). Most of the advances in global ocean tide modeling have only been made since the launch of European Remote Sensing Satellite (ERS)-1 in 1991 and Topography Experiment (TOPEX)/Poseidon (T/P) in 1992. Based on altimetry and tidal models, it is now possible to predict ocean tides globally, including in the deep ocean, with a precision of 2-4 cm over periods of several months to years (Le Provost 2001). Global information on ocean tides has resulted in an improved ability to model and predict them that would not have been possible without satellite information, due to the limitations of in situ tidal observations in the open ocean. This in itself is a major achievement. Consequently, the marine shipping sector has benefited from improved tidal predictions.

THE TURBULENT OCEAN

By providing the ability to measure the eddy variability globally, to determine its space-time variability, and to study its time evolution, altimetry led to a paradigm shift in oceanography in the late 1990s. The direct observations of the extensive eddy field by altimetry (see below) coupled with the recent focus on energy sources for internal wave mixing of the deep ocean (see next section), including those with tidal components, changed the way we think about the nature of the global ocean circulation (Wunsch and Ferrari 2004). Before altimetry the energy supply for the large-scale circulation was believed to be dominated by surface buoyancy forces related to changes in water temperature and salinity across and within the ocean basins leading to calculations and predictions of slowly changing large-scale and slow-moving features. Since the advent of altimetry, scientists know that energy is provided to the general circulation primarily by winds and tides. Perhaps the greatest single conceptual change (still not universally understood) is that the ocean is an extremely time-dependent, turbulent environment, with no steady-state patterns.

This new view of ocean dynamics has implications for understanding how the ocean has affected climate over geological time. Ocean dynamics are fundamental to understanding how heat is transferred between the ocean and atmosphere and how heat is moved from the tropics to the poles. New insights into the importance of tidal energy dissipation to ocean dynamics and to other characteristics of a turbulent ocean led to a new appreciation of the difficulty of trying to model paleoocean circulation based on proxies of scalar properties (e.g., temperature) inferred from measurements of ocean sediment cores (Wunsch 2007). A poor description of ocean circulation will lead to inaccurate models of climate change over geological time due to the high dependence of the Earth's climate on ocean circulation (Wunsch 2007). Thus, the new knowledge gained from satellite observations has the potential to greatly improve the accuracy of ocean circulation models in the future.

Internal Tides and Their Contribution to Ocean Mixing

In addition to the impressive advances in ocean tide modeling, satellite altimetry revealed how ubiquitous and important internal tides were in the open ocean. Although the importance of internal tides to the continental shelf regions has long been known, satellite observations of the open ocean tidal signal allowed scientists to calculate their significant contribution to deep ocean mixing (Garrett 2003).

This discovery not only transformed oceanography but also has major implications for climate change science.

Because internal tides result in a vertical surface displacement of only a few centimeters (a 1-cm surface elevation change corresponds to vertical displacements of isotherms of tens of meters) and are only on the order of 100 km long, early satellite altimetry measurements were not able to resolve such small variation in the sea surface height. However, since T/P, along-track analysis became possible with the availability of precise altimetry data leading to direct global measurements of internal tides (Tierney et al. 1998). Similar to tides at the ocean surface, internal tides spread as a wave within the ocean interior, and their amplitude has been shown to correlate well with features on the ocean bottom such as ridges and seamounts (Ray and Mitchum 1997). Internal tides are now considered equal to winds in generating energy for mixing.

Tides transfer 3.5 TW of energy from the Sun and the Moon to the ocean. The conventional view was that dissipation of this energy occurred on the continental shelves and was thus irrelevant to the general circulation of the ocean (Wunsch and Ferrari 2004). An unexpected finding from altimetry measurements was that internal waves of tidal period were much more prevalent and of higher amplitude than previously believed (Egbert and Ray 2000). Calculations showed that as much as 1 TW of the 3.5-TW tidal energy input could be available to mix the deep ocean (Munk and Wunsch 1998). Much of the tidal energy released in the deep ocean occurs in the presence of ocean ridges, seamounts, and other features of abyssal topography. Altimetric internal tide measurements led directly to the current physical oceanography focus on energy sources for the general circulation and the implication that both winds and tides control the circulation through mixing of the abyss. This had never even been discussed prior to about 1997.

Altimeter Measurements of Westward-Propagating Sea Surface Height Variability

From theoretical considerations, energy input to the ocean from wind and thermal forcing is expected to propagate westward in the form of Rossby waves. Rossby waves are large, slow-moving features that generally move across the ocean from east to west. Typical wavelengths are 1,000 km and longer with sea surface height (SSH) signatures of about 10 cm. While the existence of these waves had been accepted since the seminal studies by Rossby et al. (1939) and Rossby (1940), observational verification remained elusive until the accumulation of shipboard observations by the mid-1970s of a sufficiently long and spatially dense collection of vertical profiles of upper-ocean thermal structure in the North Pacific.

Satellite altimetry demonstrated the prevalence and thus the importance of Rossby wave-like variability of the ocean circulation—a central underpinning of all understanding of oceanic variability. The orbital configuration of the T/P altimeter was particularly well suited to study these features because this altimeter was specifically designed to avoid aliasing by tides. A global synthesis of T/P data by Chelton and Schlax (1996; updated by Fu and Chelton 2001) detected the expected westward propagation with latitudinally varying propagation speed in all ocean basins. Thus, T/P altimetry provided compelling evidence supporting the theory that Rossby waves are an important mechanism for moving energy from east to west in ocean basins.

A new view is evolving due to the availability of simultaneous measurements of SSH by the T/P and the European Remote Sensing Satellite (ERS) altimeters, which allows the construction of much higher-resolution SSH fields than can be obtained from a single altimeter (Chelton et al. 2007b). By merging the T/P and ERS altimeter data sets, SSH fields are obtained with approximately double the spatial resolution of SSH fields constructed from T/P alone (Ducet et al. 2000; Figure 8.5). The newly merged data set in the lower panel of Figure 8.5 shows the intricate structure of the ocean circulation. The observations of the time-dependent motions visible in this figure led to a much clearer understanding of the role such motions play in the time-varying ocean circulation. At latitudes equatorward of about 25 degrees, a Rossby wave-like character is still evident in the merged data. At higher latitudes, however, the doubling of resolution reveals that the SSH field is much more eddy-like in nature than suggested from maps constructed from only the T/P data (Chelton et al. 2007b).

Animations of the merged T/P-ERS data reveal that the resolved eddies propagate considerable distances westward. When an automated eddy-tracking procedure—developed for and applied to previous studies (Isern-Fontanet et al. 2003, 2006; Morrow et al. 2004)—is applied to the global data set, more than 8,300 eddies are trackable for 18 weeks or longer, and more than 500 eddies are trackable for more than a year (Chelton et al. 2007b). Although in a few regions there are preferences for eddy polarity, in most there is no significant difference between the numbers of cyclonic and anticyclonic eddies.

A striking characteristic of the eddy trajectories is the strong tendency for purely westward propagation. Globally, nearly 75 percent of the tracked eddies had mean propagation directions that deviated from due west by less than 10 degrees, with cyclonic and anticyclonic eddies having distinct preferences for, respectively, poleward and equatorward deflections. The fact that much of the extratropical SSH variability is attributable to nonlinear eddies rather than to linear Rossby waves (Chelton et al. 2007b)—as suggested by earlier analyses—may have significant implications for biological processes in the ocean because nonlinear eddies, in contrast to Rossby waves, transport properties vertically and horizontally.

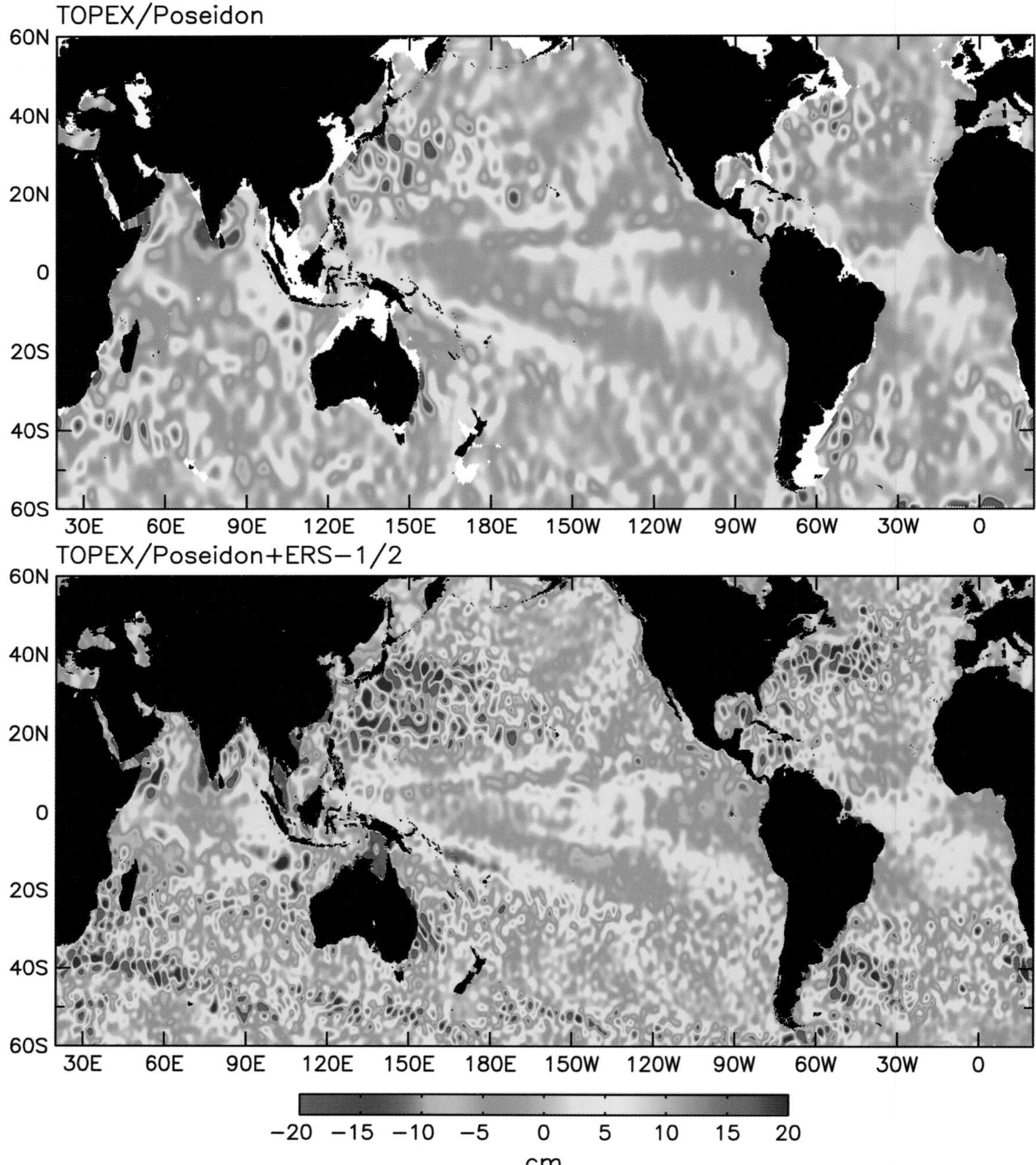

FIGURE 8.5 Global maps of SSH centered on August 28, 1996, constructed from T/P data alone (top) and from the merged T/P and ERS data (bottom). Based on the resolution limitations imposed by sampling errors (Chelton and Schlax 2003), the T/P data were smoothed with half-power filter cutoffs of 6° × 6° × 30 days, and the merged T/P-ERS data were smoothed with half-power filter cutoffs of 3° × 3° × 20 days. After filtering to remove large-scale heating and cooling effects unrelated to mesoscale variability, the anomaly SSH field consists of many isolated cyclonic and anticyclonic features (negative and positive SSH, respectively). SOURCE: Modified from Chelton et al. (2007b). Reprinted with permission by American Geophysical Union, copyright 2007.

OCEAN WIND MEASUREMENTS REVEAL TWO-WAY OCEAN-ATMOSPHERE INTERACTION

Scatterometers have also made significant contributions to the study of ocean dynamics by providing a synoptic view (approximately 25 km spatial resolution) of vector winds over the ocean. The results showed new insights into the exchange of heat and momentum between the atmosphere and ocean. Weather forecasting has been significantly improved by incorporating scatterometer-derived winds into forecasts (see Chapter 3). In particular, scatterometer data are particularly useful for determining the location, strength, and movement of cyclones over the ocean. Furthermore, new insights as to the underlying physics affecting air-sea interaction have significant implications for ocean mixing, which is important for understanding the dynamics of ocean currents as well as the supply of nutrients supporting biological productivity.

Prior to the availability of scatterometer measurements of ocean vector winds, most of what was known about the space-time variability of the wind field over the ocean was based on 10-m wind analyses from the European Centre for Medium-Range Weather Forecasting (ECMWF) and the U.S. National Centers for Environmental Prediction (NCEP) global numerical weather prediction models. Feature resolution in these models is limited to wavelength scales longer than about 500-km (Milliff et al. 2004, Chelton et al. 2006), despite the fact that winds from the QuikScat scatterometer have been assimilated into both of these models since January 2002. The resolution is even worse in the reanalysis wind fields that are used in most models of ocean circulation and for most studies of climate variability. For example, the resolution limitation of the NCEP reanalysis winds is about 1,000 km (Milliff et al. 2004). As reviewed by Kushnir (2002), ocean-atmosphere interaction on the large scales resolvable by global atmospheric models is characterized by stronger winds over colder water.

An important satellite scatterometer contribution from QuikScat data revealed that ocean-atmosphere interaction is fundamentally different on scales shorter than about 1,000 km that are poorly resolved by global atmospheric models. As reviewed by Xie (2004), low-level winds are locally stronger over warm water and weaker over cold water throughout the oceans wherever strong SST fronts exist. This ocean-atmosphere interaction apparently arises from SST modifications of stability and vertical mixing in the marine atmospheric boundary layer (MABL). This is consistent with earlier in situ studies in the Gulf Stream (Sweet et al. 1981) and the Agulhas Current (Jury and Walker 1988) that observed enhanced vertical turbulent mixing as cold air passes over warm water, deepens the MABL, and mixes momentum downward from aloft to the sea surface, thus accelerating the surface winds. Decreased mixing over cold water stabilizes and thins the MABL, resulting in decreased surface winds. Wallace et al. (1989) hypothesized a similar SST influence on low-level winds in the eastern tropical Pacific based on historical observations of surface winds and SST from ships.

The SST influence on low-level winds has important implications for both the ocean and the atmosphere. The spatial variability of the SST field in the vicinity of meandering SST fronts induces curl and divergence in the surface wind stress field that are linearly proportional to, respectively, the crosswind and downwind components of the SST gradient (Chelton et al. 2004). An example of this SST influence on the curl and divergence of the wind stress is shown in Figure 8.6 for the California Current region.

For ocean applications the wind stress curl is of particular interest because it generates open-ocean upwelling and downwelling that drive the ocean circulation and bring cold water and nutrients to the sea surface. The SST influence on the wind stress curl field results in first-order perturbations of the large-scale background wind stress curl (O'Neill et al. 2003, Chelton et al. 2007a) with timescales on the order of a month. Therefore, this ocean-atmosphere interaction likely has strong effects on both the physics and the biology of the ocean. Moreover, the feedback effects of SST-induced wind mixing and wind stress curl on the ocean alter SST, thus resulting in two-way coupling between the ocean and the atmosphere.

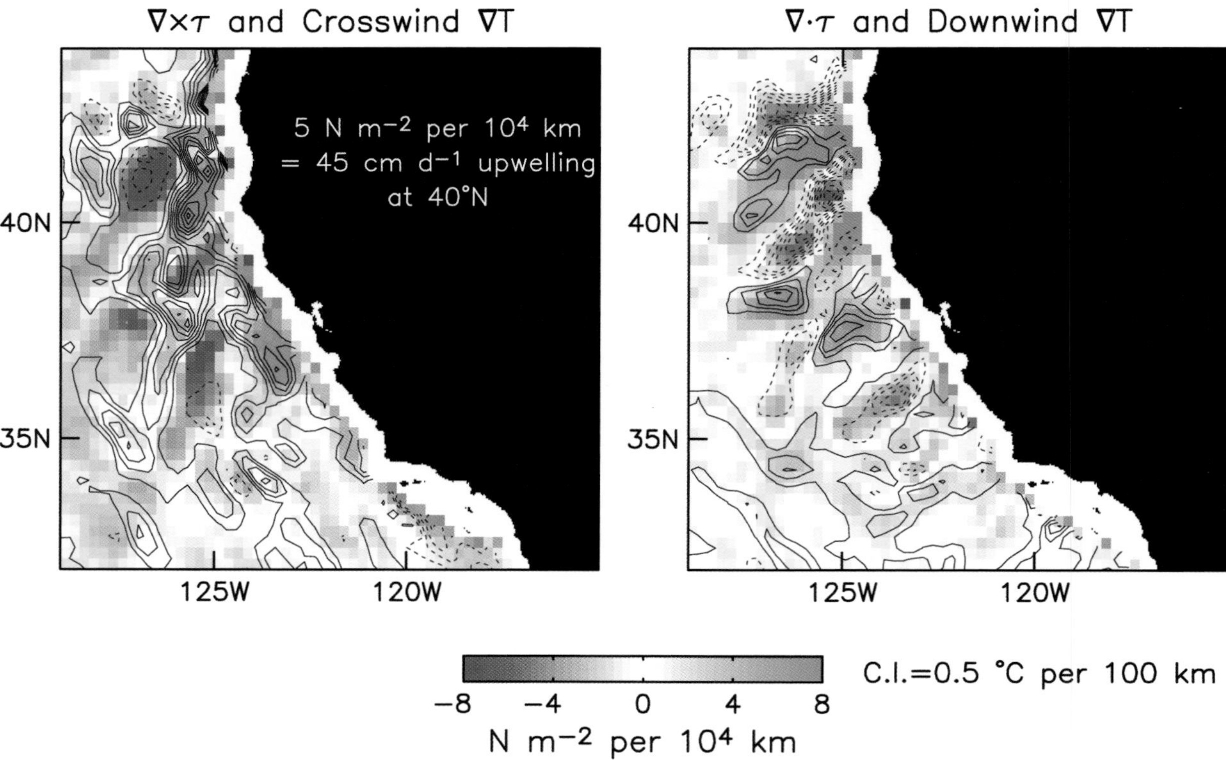

FIGURE 8.6 September 2004 averages of wind stress curl with contours of crosswind SST gradient (left) and wind stress divergence with contours of downwind SST gradient (right) over the California Current system. The wind stress fields were constructed from QuikScatdata. The SST fields were constructed from the U.S. Navy Coupled Ocean/Atmosphere Mesoscale Prediction System (COAMPS). Satellite microwave measurements of SST are not well suited to studies in this region because of the coarse (~50 km) resolution and the inability to measure SST closer than ~75 km to land. SOURCE: Chelton et al. (2007a). Reprinted with permission from the American Meteorological Society, copyright 2007.

9

Ecosystems and the Carbon Cycle

Research on the biosphere aims to understand and predict how terrestrial and marine ecosystems are changing, how they are affected by human activity or through their own intrinsic biological dynamics, how they respond to climate variations, and in turn how they affect climate. One of the primary goals of ecosystem research is to determine the amount of primary production, which is most commonly expressed in units of carbon incorporated during photosynthesis and estimates the amount of energy available for higher trophic levels. Since the discovery of the importance of carbon dioxide as a greenhouse gas, the estimation of global carbon fixed by photosynthetic processes has become a central quest in global carbon cycle research and an integral part of climate models.

Before the satellite era, few scientists had attempted to estimate these parameters at a global scale. Instead, most research efforts were dedicated to understanding local dynamics because ecosystem processes are highly variable in response to localized environmental changes. Orbiting satellites provide an ideal vantage point for viewing dynamic ecosystems on the land and in the ocean (Box 9.1). This chapter discusses how the remarkable technological advances of the past decades have enabled scientists to compose routinely global maps of terrestrial and marine productivity, assess the role of the ocean in the global carbon cycle, observe long-term ecosystem trends and atmosphere-biosphere coupling, and even study plant physiology from space.

For the first time, remote sensing made direct global observations of photosynthesis, plant growth, and ecosystem phenology possible, leading to the evolution of a global perspective on ecology (Boxes 9.2 and 9.3). Charles Keeling's continuous measurements of atmospheric carbon dioxide (CO_2) concentrations at Mauna Loa, beginning in 1957, showed a seasonal signal in the atmospheric CO_2 concentration due to the terrestrial biosphere being a source and sink for carbon during the winter and summer, respectively. Subsequent work showed that atmospheric CO_2 was steadily increasing (Keeling et al. 1976) and that it stemmed from fossil fuel burning, catalyzing an interest in obtaining a global perspective of the carbon cycle. In 1982, the National Aeronautics and Space Administration (NASA) held a workshop in Woods Hole, Massachusetts, on global change (Goody 1982) that spurred a subsequent paper by Tilford (1984) presenting the scientific rationale for the Earth Observing System (EOS). These papers called attention to how anthropogenic global changes might impact ecosystems.

TERRESTRIAL PRIMARY PRODUCTIVITY

New awareness of the relationship between microclimate and plant functions in the 1970s and 1980s spurred the development and evolution of field-portable instruments to measure plant physiological processes, photosynthesis, and transpiration, moving these measurements from the laboratory to the field. Despite these newly available field instruments, global observations of ecosystem and larger-scale processes did not become available until the advent of satellite observations because the field measurements were generally restricted to short-term (seconds to minutes) leaf measurements. The capability of assessing plant productivity from satellite radiance measurements (Box 9.2) opened an entirely new front in ecosystem research. Because small-scale point measurements did not lend themselves well to interpolating and creating global maps, synoptic satellite data provided the first direct globally distributed measurements of terrestrial functioning.

The NASA EOS program brought new capabilities for monitoring terrestrial productivity, with near-daily global coverage of a more capable well-calibrated Moderate Resolution Imaging Spectroradiometer (MODIS) that has allowed development of new biophysical measurements with less reliance on simple empirical indices. One of the new products is the direct global measurement at 1-km resolution of leaf area index (LAI)—an important structural property of

BOX 9.1
Ecosystems as Seen from Space

Satellite-based studies of the land and ocean ecosystems rely primarily on imaging sensors measuring radiance in the visible and near infrared. These spectral bands were ideally suited to monitor plant biomass and primary production because the chlorophyll *a* pigment, found in all marine and terrestrial photosynthetic plants, reflects green light while absorbing in the blue and red spectral regions. Because plant leaves contain no molecules with high absorption in the near infrared, they are highly reflective in this region. Therefore, the "greenness" of terrestrial ecosystems can be mapped by employing the ratio of red to infrared bands. However, this ratio does not work for the ocean because water is such a strong absorber in the red and infrared that little or no radiation is reflected out of the ocean at those wavelengths. Instead, the ratio of blue to green bands, after correcting for the atmosphere, has been used to quantify the chlorophyll concentration in the ocean (Box 9.3).

Remote sensing techniques for mapping and studying terrestrial and marine ecosystems have evolved along different paths because of different technological requirements. Compared to the ocean, the land is a bright surface whose features have distinct spectral signatures and generally sharp boundaries. The spatial scale of such features is on the order of tens of meters, thus requiring high spatial resolution, but the features generally change slowly over seasons or longer. In contrast, the ocean is a dark surface with subtle spectral variation that requires high radiometric sensitivity. Reflectance from the atmosphere dominates the signal received by a satellite over the ocean, and this signal must be estimated and removed before the ocean signal can be analyzed. Features in the ocean have spatial scales on the order of tens of kilometers, with fluid boundaries that change on timescales of hours to days. These differences have led to different sensor and mission requirements, but the goals remain similar. Both terrestrial and marine studies have sought to quantify primary productivity and the role of the biosphere in the global carbon cycle.

BOX 9.2
Converting Radiance to Plant Productivity

Jordan (1969) was the first to use a ratio of near-infrared and red radiation to estimate biomass and leaf area index (leaf area/ground surface area) in a forest understory. This study was quickly followed by application of near-infrared/red ratios to estimate biomass in rangelands (e.g., Pearson and Miller 1972; Rouse et al. 1973, 1974; Maxwell 1976) and was extended by Carneggie et al. (1974) to the Earth Resources Technology Satellite (ERTS-1) observations of seasonal growth, which showed that the seasonal peak in the near-infrared/red ratio coincided with maximum foliage production, thus effectively tracking the phenological cycle.

Rouse et al. (1974) introduced a spectral index, a normalized ratio that reduced illumination differences and other extrinsic effects by dividing the difference of the two bands by their sum, a ratio adopted as the normalized difference vegetation index (NDVI). A landmark paper by Tucker (1979) established linear relationships between vegetation spectral indices (ratios of visible and near-infrared bands) to leaf area and biomass. Following this paper, vegetation indices rapidly became an established method for analysis of plant biophysical properties using laboratory, field, airborne, and Landsat data. Today, nearly 2,000 papers have been published using the NDVI, and nearly 6,000 have used some type of vegetation index to study vegetation. These early studies established that red and near-infrared satellite bands could track changes in plant growth and development.

the plant canopy used to estimate functional process rates of energy and mass exchange, specifically to calculate rates of photosynthesis, evapotranspiration, and respiration (Figure 9.1). For the first time this measurement provides a consistent observational basis to estimate and monitor global productivity. Time series of LAI allow comparison of phenological patterns among six global terrestrial biome types. LAI is defined as the one-sided leaf area per unit of ground area and is produced by R.B. Myneni, Boston University. An algorithm is used to convert red and near-infrared band reflectances to global maps of LAI with modifications for the six biome types, taking into account the directional Sun and

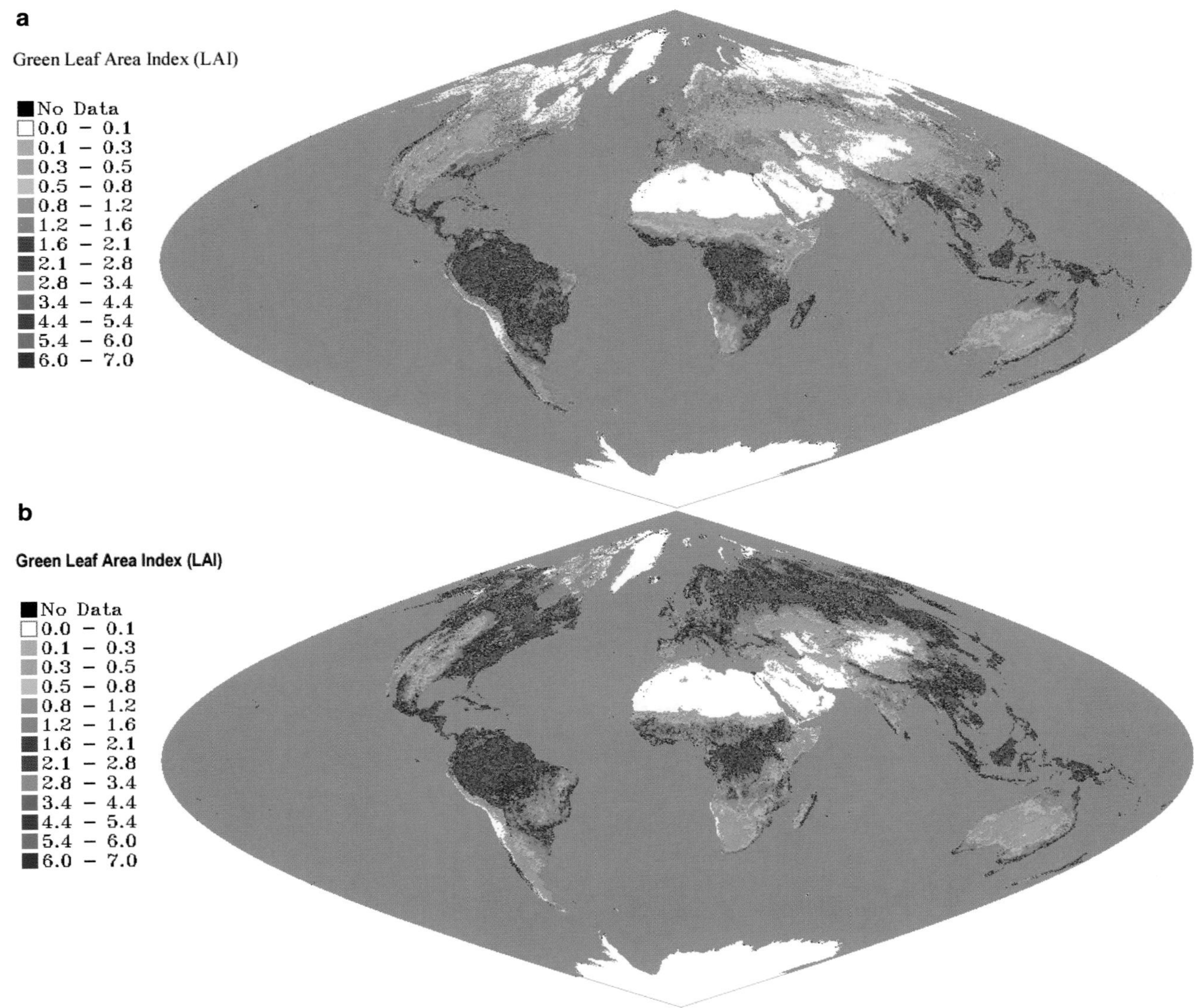

FIGURE 9.1 February (top panel) and August (bottom panel) 2006 global 4-km monthly composites of leaf area index, computed from the Moderate Resolution Imaging Spectroradiometer (MODIS; Mod15, collection 4). SOURCE: R.B. Myneni, Boston University, *http://diveg.bu.edu*.

view factors and measurement uncertainties. Prior to today's satellites, this key biophysical variable was painstakingly evaluated at the scale of small field sites by dropping a pin or line through the canopy and counting the number of leaves that were contacted. With the development of red and near-infrared indices such as normalized difference vegetation index (NDVI) in the 1980s, it became possible to correlate these ground measurements with index values, allowing the extension of direct measurements to larger regions.

Today, with MODIS, this observation has become more precise by its extension to a biophysical measurement. Satellite monitoring of the dynamics of Earth's vegetation is essential to understanding global ecosystem functioning and response to climate variability and climate change. This new observational perspective has led ecologists to see ecosystem processes in an integrated temporal and global context.

MARINE PRIMARY PRODUCTIVITY

Approximately half of all global primary production occurs in the ocean, almost entirely due to microscopic single-cell algae known as phytoplankton. In the presence of ample sunlight and nutrients, phytoplankton reproduce rapidly and biomass can double in a day. As the cells grow and reproduce, carbon dioxide dissolved in the surface ocean is converted to organic matter, which is then consumed by

BOX 9.3
Global Marine Biomass from Ocean Color Remote Sensing

The ability to derive global maps of chlorophyll *a* concentration (milligrams per cubic meter) in the upper ocean from ocean color sensors was a groundbreaking achievement for the oceanographic community (Figure 9.2). This biomass estimate can then be related to primary productivity and the marine carbon cycle. Although clouds prevent ocean color sensors to see the entire ocean surface on each orbital pass, a global picture of the distribution of photosynthetic plant biomass emerges from averaging data over several consecutive days or weeks.

The first ocean color sensor was the Coastal Zone Color Scanner (CZCS), an experimental proof-of-concept mission operating on the Nimbus 7 satellite between 1978 and 1986. The CZCS demonstrated that it is possible to detect subtle changes in the color of the ocean and relate these to the concentration of chlorophyll *a*, the light-harvesting pigment found in all plants. In particular, chlorophyll *a* concentrations are quantified by empirical algorithms relating spectral band ratios (blue to green) to the concentration of chlorophyll in the ocean (Clark 1981, Gordon and Morel 1983, O'Reilly et al. 2000). A major requirement is that the spectral radiance measurements made by the satellite be corrected to remove the effect of the atmosphere, which comprises more than 90 percent of the top-of-atmosphere signal. This was a major technological breakthrough after the launch of the CZCS (Gordon et al. 1980). Contrary to its name, the sensor was better at estimating biomass in the open ocean than in the coastal zone. Phytoplankton and dissolved organic matter are the primary sources of optical variability in the open ocean (so-called Case 1 waters [Morel and Prieur 1977, Gordon and Morel 1983, Siegel et al. 2002]), whereas in coastal regions, mixtures of organic and inorganic materials affect the ocean color. The problem of differentiating and quantifying individual constituent concentrations in the coastal ocean remains a challenge today.

The ocean color technology pioneered by the CZCS has since been improved and incorporated into modern space instruments. The first modern global ocean color sensor was Japan's Ocean Color and Temperature Sensor (OCTS) launched in August 1996 aboard the Advanced Earth Observing Satellite (ADEOS). The U.S. Sea-Viewing Wide Field-of-View Sensor (SeaWiFS) followed in August 1997, shortly after the ADEOS experienced structural damage after only 9 months in orbit. SeaWiFS is owned by Orbital Sciences Corporation, with a guarantee from NASA to buy data for the scientific research community. Ocean color data continue to be acquired by the Moderate Resolution Imaging Spectroradiometers (MODIS) aboard the Terra and Aqua satellites launched in 1999 and 2002, respectively, and by a number of other ocean color instruments operated by other countries (Table 9.1).

FIGURE 9.2 Map of chlorophyll *a* concentration (milligrams per cubic meter) in the upper Atlantic Ocean derived from data obtained by the Sea-viewing Wide Field-of-view Sensor (SeaWiFS). SOURCE: SeaWiFS Project, NASA Goddard Space Flight Center, and GeoEye.

continued

BOX 9.3 Continued

TABLE 9.1 Past and Present Satellite Sensors with Ocean Color Capability

Sensor	Satellite	Country	Dates	Spatial Resolution (m)	No. of Bands	Comments
CZCS	Nimbus-7	United States	November 1978-July 1986	825	5	Proof-of-concept
MOS	IRS P3	Germany-India	March 1996-March 2004	523	13	Requires ground station
OCTS	ADEOS-1	Japan	August 1996-June 1997	700	8	+ Four thermal IR bands for SST
SeaWiFS	SeaStar	United States	August 1997-present	1,100	8	Commercial data, free to researchers
OCI	ROcSAT-1	Taiwan	December 1998-July 2004	800	6	Latitude coverage 35° N-35° S
OCM	OceanSat-1 (IRS P4)	India	May 1999-present	360	8	+ Scanning microwave-SST
MODIS	Terra Aqua	United States	December 1999-May 2002	1,000	9	+ 27 other bands for land, atmosphere, SST
MERIS	Envisat	EU	March 2002-present	250 LAC 1,000 GAC	15	LAC data require ground station

NOTE: Nations such as Japan, Taiwan, and India have invested in ocean color as a valuable source of information for their fishing fleets (Laurs et al. 1984, Butler et al. 1988). This has also met with some success in the United States where it has been argued that the satellite information makes fishing more efficient, thus saving fuel and other limiting resources.

CZCS = Coastal Zone Color Scanner; GAC = Global Area Coverage; LAC = Local Area Coverage; MERIS = Medium Resolution Imaging Spectrometer; MODIS = Moderate Resolution Imaging Spectroradiometer; MOS = Maritime Observation Satellite; OCI = Ocean Color Imager; OCTS = Ocean Color and Temperature Sensor; SeaWiFS = Sea-Viewing Wide Field-of-View Sensor; IRS = Indian Remote Sensing Satellite; ADEOS = Advanced Earth Observing Satellite; SST = Sea Surface Temperature.

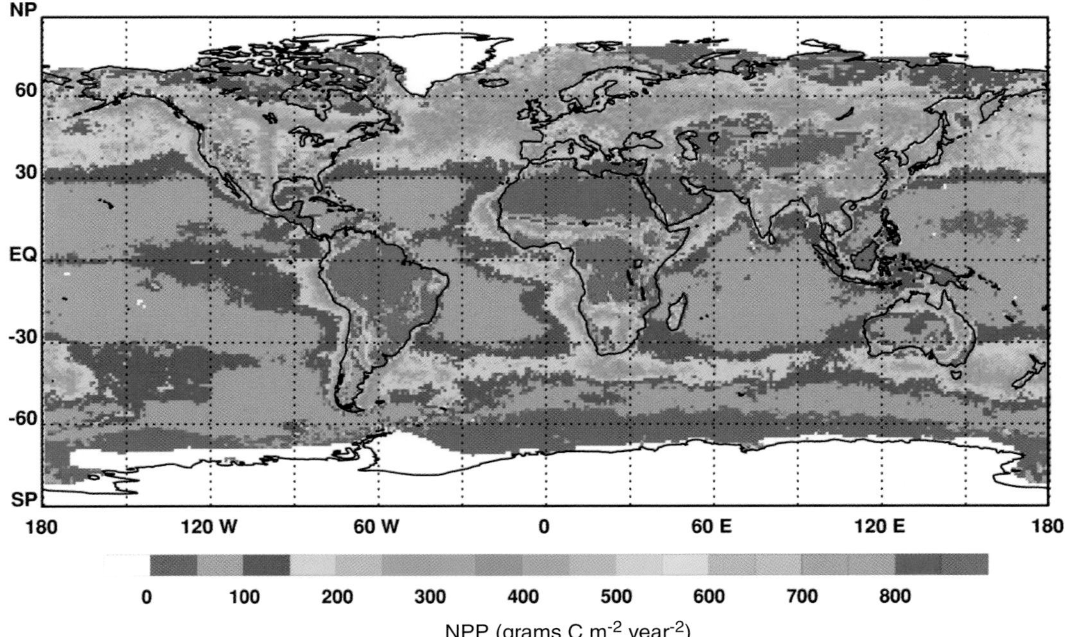

FIGURE 9.3 Global annual NPP (in grams of carbon per square meter per year) for the biosphere, calculated from the integrated CASA-VGPM (Vertically Generalized Production Model) model. Input data for ocean color from CZCS sensor are averages from 1978 to 1983. The land vegetation index from the AVHRR sensors is the average from 1982 to 1990. SOURCE: Field et al. (1998). Reprinted with permission from AAAS, copyright 1998.

zooplankton, fish, and other animals in the "food chain." Because of its rapid growth and many consumers, phytoplankton biomass or chlorophyll concentration varies on short timescales, yet the extent of a "patch" of accumulated biomass is on the order of 10-100 km.

Satellites have allowed scientists to routinely estimate phytoplankton productivity on an annual basis for the first time, enabling them to detect a trend in decreasing phytoplankton productivity associated with warming of the surface ocean at mid- to low latitudes. Because a phytoplankton bloom and its associated productivity are such large-scale yet short-lived phenomena, there is simply no way to survey large enough areas of the ocean to capture their dynamics using ships to map phytoplankton biomass and productivity.

Prior to the introduction of satellite observations, estimates of oceanic primary production depended on relatively few labor-intensive ship-based incubations using the ^{14}C technique that had become the standard method for measuring primary productivity in the ocean (Steeman-Nielsen and Jensen 1957). To estimate global annual oceanic production (gigatons of carbon per year), the mean integral productivity was first estimated for the different oceans and depth ranges using relatively few measurements made in each domain. These were then multiplied by the area of the ocean domain and 365 days per year to derive annual oceanic primary production. Due to the vastness of the ocean and high spatial and temporal variability, ship-based global mapping was infrequently attempted and could not realistically capture the interannual variability. Even with the development of fluorescence-based estimates of marine primary productivity, which could be obtained from instruments towed behind ships, obtaining global coverage would still require years. It has long been recognized that ship-based sampling methods suffer from significant undersampling in both space and time (McCarthy 1999). Consequently, the best quantitative global estimates of both biomass and productivity are derived with the use of satellite observations that provide the necessary frequency of global coverage.

To estimate primary productivity from satellite measurements, it is assumed that the productivity is proportional to the phytoplankton biomass. Consequently, measuring biomass is the first critical step in estimating marine primary productivity from space. Chlorophyll *a*, the ubiquitous light-harvesting pigment found in all green plants, has long been a standard measure of phytoplankton biomass (Box 9.3, Figure 9.2, Table 9.1). This is largely because chlorophyll can be measured rapidly and easily owing to its fluorescent and absorption properties.

Early estimates of oceanic primary productivity derived using satellite data provided a relative static picture in that they represented average annual productivity (Platt and Sathyendranath 1988, Antoine et al. 1996, Behrenfeld and Falkowski 1997, Field et al. 1998). One of the most thorough estimates was that of Longhurst el al. (1995), who estimated global ocean net primary production using Coastal Zone Color Scanner (CZCS) data and models of the subsurface chlorophyll distribution and Photosynthesis-irradiance (*P-I*) relationships defined for 57 biogeochemical provinces.

GLOBAL MARINE AND TERRESTRIAL PRIMARY PRODUCTION

Net primary productivity (NPP) is influenced by climate and biotic controls that interact with each other. Field et al. (1995) predicted global terrestrial NPP on a monthly time step using the Carnegie-Ames-Stanford Assimilation (CASA) model, incorporating a set of ecological principles and satellite and surface data. Several authors have used satellite data to estimate global net primary production, combining both terrestrial and oceanic models. Within a few years they used a linked ocean-terrestrial model that combined an 8-year Advanced Very High Resolution Radiometer (AVHRR) record and a 6-year CZCS data record with a biogeochemistry model to estimate global land and ocean NPP (Field et al. 1998, Figure 9.3). This study found that the contribution of land and ocean to NPP was nearly equal but that there was striking variability in NPP at a local level. Based on the spatial variability in the satellite data, their model predicted strong differential resource limitations for terrestrial and ocean habitats.

Behrenfeld et al. (2001) used the Sea-Viewing Wide Field-of-view Sensor (SeaWiFS) data to estimate terrestrial and ocean primary production during the transition between El Niño and La Niña conditions in 1997 to 1999. They found that the ocean exhibited the greatest effect, particularly in tropical regions where El Niño-Southern Oscillation (ENSO) impacts on upwelling and nutrient availability were greatest. Terrestrial ecosystems did not exhibit a clear ENSO response, although regional changes were substantial. These studies clearly demonstrate the invaluable contribution satellite observation of NPP make to the fundamental understanding of climate change impacts on the biosphere.

THE OCEAN CARBON CYCLE

Satellite observations afford the only means of estimating and monitoring the role of ocean biomass as a sink for carbon. In particular, the fundamental question of whether the biological carbon uptake is changing in response to climate change can only be addressed with satellite measurements. It requires not only ocean color measurements (phytoplankton biomass and productivity) but also coincident space-based observations of the physical ocean environment (circulation and mixing) and land-ocean exchanges through rivers and tidal wetlands, as well as winds, tides, and solar energy input to the upper ocean. Observing linkages between the physical and chemical environment and the biology of the ocean is a significant achievement of observations from space. Continuity of this record is critical. Understanding the consequences of the CO_2 increase and its effect on terrestrial and marine ecosystems will require global-scale long-term observations from carefully calibrated satelliteborne sensors.

Early carbon cycle models that were used to investigate sources and sinks of anthropogenic CO_2 ignored the effects of marine productivity, which was thought to be in equilibrium on annual timescales. Since marine productivity is not limited by carbon, it was reasoned that increases in CO_2 would not affect oceanic productivity. More recently, modelers have investigated how marine productivity might be affected indirectly by climate change through its effect on oceanic and atmospheric circulation patterns.

Because phytoplankton life cycles are orders of magnitude shorter (days versus years or decades) than those of terrestrial plants, phytoplankton may respond to climate influences on ocean circulation, mixing, and the supply of nutrients and light much more quickly than plants in terrestrial ecosystems. Given that oceanic primary productivity is estimated to be roughly half of all global primary productivity, the oceanic component of the carbon cycle will respond more quickly to climate changes.

For example, there are vast areas of the Pacific and Southern Oceans, where phytoplankton productivity might be limited by iron (Martin et al. 1994). In contrast to the other limiting nutrients, which are supplied primarily by the deep ocean, atmospheric dust deposition is one of the main sources of iron to the open ocean. Paleorecords indicate that the Southern Ocean responded with increased productivity during colder periods when iron atmospheric deposition was enhanced due to the expansion of arid regions. This led to the notion that these areas in the Pacific and Southern Oceans could be stimulated to draw down large amounts of atmospheric CO_2 if they were provided with iron. Several experiments conducted in the late 1990s and early 2000s proved conclusively that iron does limit production in these regions (Coale et al. 2004). Iron is supplied to the open ocean by atmospheric transport (dust deposition), by lateral advection of waters from the continental margins, and by upwelling of deep iron-rich waters. Long-term monitoring of the ocean phytoplankton will reveal whether climate change will affect these iron supplies potentially fertilizing the Southern Ocean or the Pacific.

With 10 years of continuous ocean color data (since 1997), we now have the ability to observe year-to-year variability in global oceanic primary production and begin to assess longer-term trends in ocean carbon uptake. Behrenfeld et al. (2006) describe a steady climate-driven decrease in oceanic NPP related to the warming of permanently stratified ocean waters at mid- to low latitudes over the past 8 years. This period of decreasing NPP followed the rise in NPP between the El Niño and La Niña phases. Satellite observations afford the only means of estimating and monitoring the role of the ocean biomass as a sink for carbon.

LONG-TERM ECOSYSTEM RECORD REVEALS ATMOSPHERE-BIOSPHERE COUPLING

Although early studies established that red and near-infrared satellite bands could track changes in plant growth and development (Box 9.1), the large number of Landsat images (~5,000) required to assemble a global database, combined with computational requirements and frequent cloud cover, have prevented analysis of complete global or time series of Landsat data sets. Launched in 1978, the Coastal Zone Color Scanner showed that ocean productivity could be observed using visible and near-infrared bands; however, CZCS measurements were saturated over land and thus unusable.

The Advanced Very High Resolution Radiometer on the National Oceanic and Atmospheric Administration's (NOAA) polar-orbiting weather satellites has obtained a continuous record of daily global observations since 1978, acquiring both red and near-infrared bands. Because AVHRR was not designed for observing the terrestrial biosphere and the 1- to 8-km scale of AVHRR pixels was significantly larger than theoretical understanding of ecosystem processes, scientists were initially skeptical about whether biospheric patterns and trends could be observed. However, scientists have managed to overcome technical problems such as maintaining calibrations, screening clouds, and adjusting for different observational angles. Thanks to the pioneering efforts of Compton Tucker, the daily AVHRR data set now spans more than 25 years and is the longest continuous global record available of terrestrial productivity, phenology, and ecosystem change for monitoring biospheric responses to climate change and variability. Although AVHRR was not designed for climate monitoring, continuing improvements in calibration and reanalysis have produced a consistent record for monitoring and assessing past and future biospheric responses resulting from climate change and variability and anthropogenic activities.

Initial studies using AVHRR followed seasonal and annual trends in ecosystem production and vegetation phenology at regional and continental scales (Tucker et al. 1985, Townshend et al. 1985) and at the global scale (Justice et al. 1985). In the early 1990s some key papers introduced the use of remote sensing data to ecology (Roughgarden et al. 1991, Ustin et al. 1991) and stressed the need for ecologists to focus on global ecological problems (Mooney 1991). These ideas led to the resurgence in ecosystem research and modeling of biogeochemical processes and significant advances in understanding the Earth as a system.

By the mid-1990s, global ecosystem and biogeochemical models used satellite data to establish variable vegetation composition and abundance (e.g., Biome BioGeochemical Cycles [BGC], Running and Hunt 1993; CASA, Potter et al. 1993). The concept of resource limitations as the controlling mechanism determining NPP was established in the late 1980s (Chapin et al. 1987). This placed a premium on direct satellite observations of vegetation conditions to provide more realistic estimates of NPP. Previous estimates used uniform rates of NPP for each land-cover type and assumed that NPP is proportional to reflected net shortwave radiation.

The relationship between vegetation indices and the physiological processes of photosynthesis and absorbed photosynthetic radiation (APAR) were formalized in theoretical analyses by Piers Sellers (1986). These developments led to a seminal paper by Tucker et al. (1986) in which it was shown that changes in the planetary NDVI (greenness) were strongly correlated with daily dynamics of terrestrial IPAR (intercepted photosynthetically active radiation) and atmospheric CO_2 concentrations. There is a strong negative correlation between NDVI and atmospheric CO_2 such that NDVI is high when CO_2 concentrations are low and low when CO_2 concentrations are high (Figure 9.4). This temporal pattern in ecosystem photosynthesis and respiration demonstrates the dynamic coupling between the biosphere and the atmosphere.

In the past decade, NDVI data from AVHRR have become a critical component in monitoring climate change (Fung et al. 1987, Sellers et al. 1994, Angert et al. 2005), assessing changing length and timing of the growing season (e.g., Justice et al. 1985, Myneni et al. 1997, 1998; Box 9.4, and Figure 9.5), and monitoring the state of the biosphere (Anyamba et al. 2001) and other ecosystem phenomena. Long-term records of NDVI have revealed its increase in response to a warming climate during the 1980s and early 1990s, but this trend has leveled off most recently (Angert et al. 2005).

STUDYING PLANT PHYSIOLOGY FROM SPACE

To estimate actual NPP in the presence of environmental stressors, researchers developed methods to remotely estimate regulatory plant biochemicals. The first advance was the development of the "photochemical reflectance index" (PRI) by John Gamon and colleagues (Gamon et al. 1992) to better predict radiation use efficiency. This index has had extensive use for noninvasive studies of leaf photosynthesis by plant physiologists, although at the image level it appears more related to carotenoid content. The PRI has led to a range of other studies to quantify plant pigments and develop methods for assessing them. These advances follow increasingly specific knowledge of spectroscopy of plant properties and how this information can be retrieved from satellite sensors.

A radiative transfer model, developed by Jacquemoud

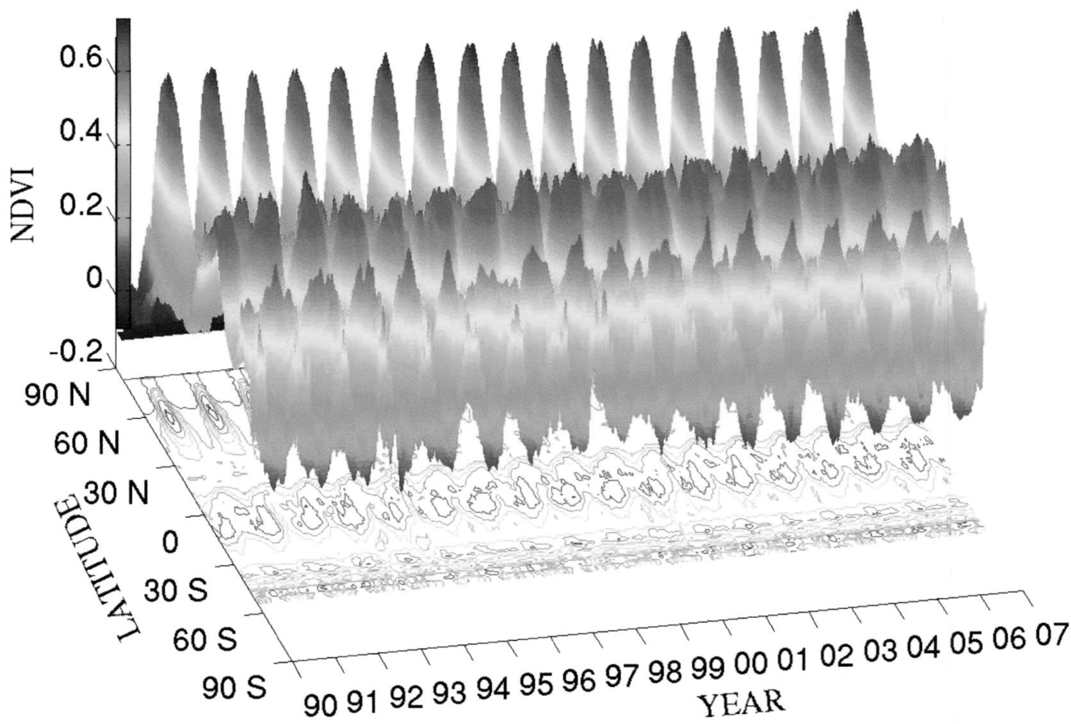

FIGURE 9.4 Weighted NDVI data plotted against time and latitude zone. Note the highly seasonal effects in the northern latitudes, the influence of deserts in the 20°-30° N latitude zone, the generally constant response in equatorial areas, and the influence of the low proportion of land area south of 30° S. SOURCE: Reprinted with permission from J.E. Pinzon (SSAI-NASA/GSFC) and C.J. Tucker (NASA/GSFC).

BOX 9.4
Increasing Growing Season

Myneni et al. (1997) published a groundbreaking paper using daily satellite data over a 9-year period to show increases in the length of the growing season in the boreal region. They used a time series of NDVI, a measure of the photosynthetic activity of vegetation canopies, derived from the daily AVHRR satellite data, and showed an increase in length of the growing season in the boreal region (north of 45°) of 12 days (8 days in spring and 4 days in autumn) from 1981 to 1991. They demonstrated that this extension of the growing season and enhanced amplitude of NDVI over the summer were likely correlated with warmer spring and autumn temperatures over the region. This result partially corroborated an estimated 7-day extension of the growing season that was inferred from atmospheric CO_2 measurements. Uniquely, their analysis detected significant spatial variation in the distribution of enhanced NDVI, with western and eastern Canada and southern and central Alaska having large increases in contrast with little change in other areas, such as central Canada and Siberia. Monitoring the spatially variable increase in biospheric activity over the circumpolar region was only possible because of the availability of polar-orbiting satellites.

Furthermore, scatterometer data from satellites provide further evidence that the growing season has lengthened in the Arctic region over the past 20 years. Figure 9.5 shows the progression of the spring 2000 thaw in Alaska. Similar measurements made since 1988 show that the thaw in the Arctic has been advancing by almost 1 day a year. These observations could not have been made without satellites since melting occurs rapidly across the Arctic during the period of melt and the timing varies between years, depending on weather conditions.

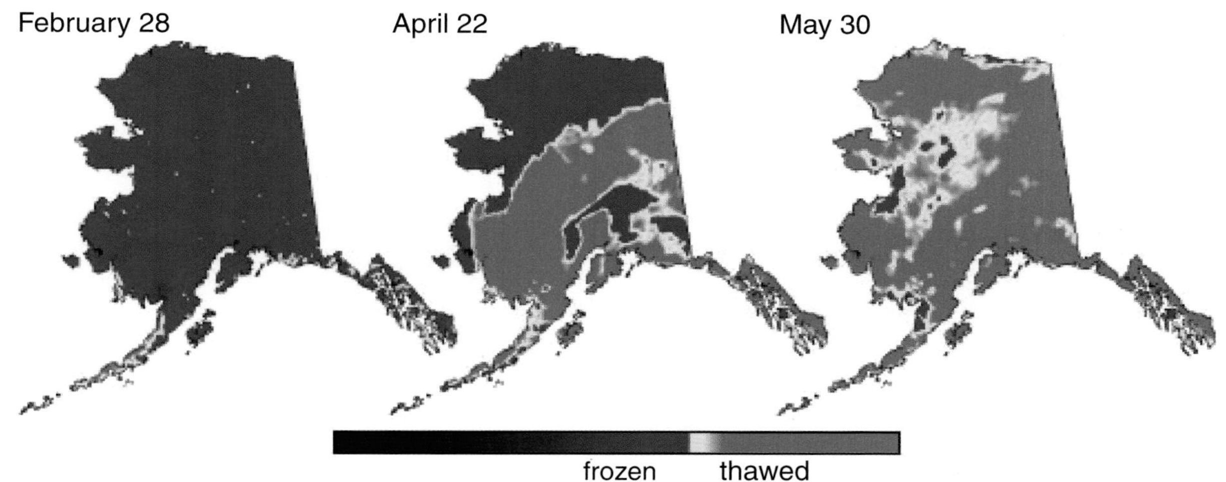

FIGURE 9.5 Progression of the spring thaw in Alaska during the year 2000 with snow and ice (blue), ice and slush with bare ground (yellow), and water and bare ground (red). A series of SeaWinds scatterometer measurements on the QuickScat satellite, which are sensitive to water in frozen and liquid states, were used to make these images. SOURCE: Kimball et al. (2006). Reprinted with permission from the American Meteorological Society, the American Geophysical Union, and the Association of American Geographers, copyright 2006.

and Baret (1990), has rigorously demonstrated the potential to retrieve several plant biochemicals from reflectance and transmittance data and is in wide use today. As summarized by Ustin et al. (2004), the list of plant biochemicals has become longer with studies of chlorophyll fluorescence (Zarco-Tejeda et al. 2000a, b), canopy water content (Gao and Goetz 1995, Zarco-Tejeda et al. 2003), and canopy nitrogen content (Kokaly and Clark 1999). Many of the more recent advances are based on new imaging spectroscopy technology using NASA's Airborne Visible/Infrared Imaging Spectrometer (AVIRIS), an aircraft instrument operated by the Jet Propulsion Laboratory since 1987. NASA has flown one hyperspectral imager in space, the Earth Observing-1 Hyperion, which was launched in 2000 as an engineering test bed yet has continued to operate to today. This technology has significant promise for continued advances in detecting biochemical properties of interest but also for using the high dimensionality of the data to improve land-cover and land-use classifications. Several recent studies have used NASA's AVIRIS and other hyperspectral imagers to map invasive weeds with high specificity (Figures 9.6 and 9.7; see also Box 9.5, Williams and Hunt 2002, Underwood et al. 2003, Asner and Vitousek 2005).

In a span of slightly more than 25 years, NASA instruments and the research supported by the agency have evolved from primitive correlative studies to physically based accurate analyses. Understanding has advanced rapidly with the synergistic advent of new sensor capabilities such as increased signal to noise ratios, with simultaneous higher spatial and spectral resolution, radiometrically stable instruments, accurate geolocation of images due to advances in satellite pointing control and Global Positioning System (GPS), and development of atmospheric radiative transfer models allowing retrieval of accurate reflectance data. Computer advances have allowed more complex analytical methods to be developed that better match the spatial and spectral patterns in the data. The extensive research funded by NASA through the Earth Observing System program and the scientific advances in understanding our home planet over the past two decades represent a major achievement of the space program.

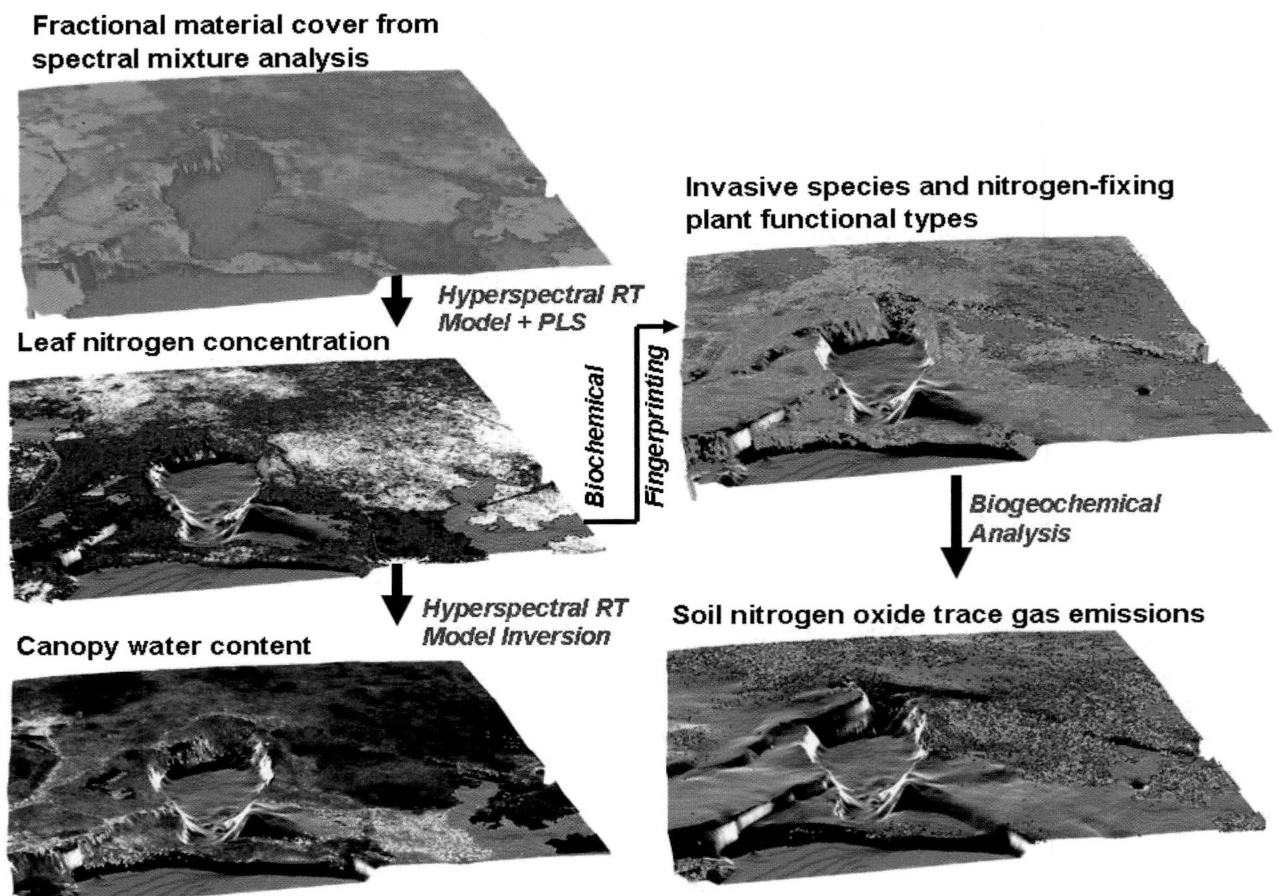

FIGURE 9.6 A map of invasive species in the Hawaiian rainforest, measured using NASA's AVIRIS data and impacts of invasive species and plant functional types on biogeochemical cycles. SOURCE: Modified from Asner and Vitousek (2005).

BOX 9.5
Detecting Invasive Plant Species

All global ecosystems, with the possible exception of Antarctica, are impacted by invasive species that are substantially changing their functional and structural integrity. Invasion of natural ecosystems represents a serious threat to global biodiversity. Factors attributed to the spread of these species include climate change, land use, land conversion, resource extraction, and habitat fragmentation, combined with international transport. Substantial economic costs are associated with these changes, from loss of agricultural production and increased wildfire frequency to loss of recreational potential. Costs in the United States alone are estimated to exceed $120 billion per year (Pimentel et al. 2005). Recent advances in imaging spectroscopy, a technique to measure a detailed spectrum for all pixels in the image have allowed mapping of individual species and plant communities based on their spectral characteristics. Underwood et al. (2003) used this data to map invasive species in native shrublands along the central coast of California at Vandenberg Air Force Base. Figure 9.7 shows the distribution of invasive species and native plant communities at 3-m pixel resolution for part of the base along the Pacific Coast shoreline. This information is being used by land managers to improve efficiencies in eradication and containment programs. Data of the quality required for mapping individual plant species must currently be acquired by airborne hyperspectral imagers. NASA's suborbital sciences program has led to the development of this cutting-edge technology and has supported the research required to use it effectively, as shown in the figure.

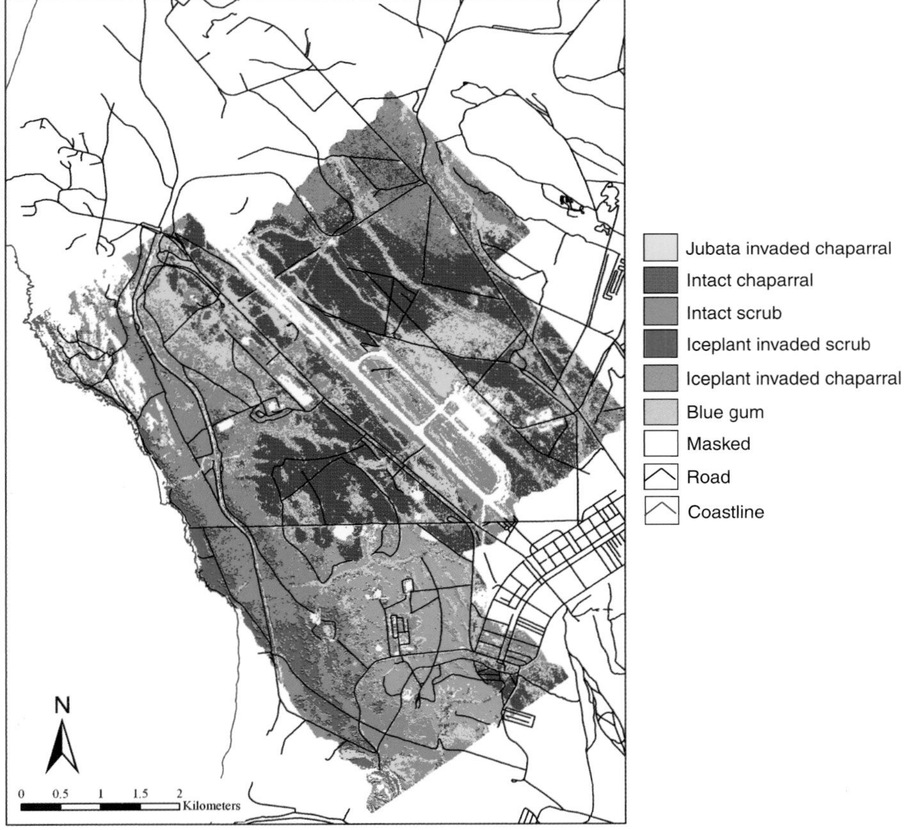

FIGURE 9.7 Distribution of three invasive species—iceplant, jubata grass, and blue gum—in two native shrub ecosystems—coastal sage scrub and Burton Mesa chaparral—on the central coast of California. The map was produced from a mosaic of flightlines acquired from airborne NASA AVIRIS data, a 224-band imaging spectrometer measuring from the visible through the solar infrared (400-2,500 nm) and measured at a nominal 3-m pixel resolution. SOURCE: Underwood et al. (2006). Reprinted with kind permission of Springer Science and Business Media, copyright 2006.

10

Land-Use and Land-Cover Change

Human activities are transforming Earth's surface at unprecedented rates by ubiquitous exploitation of Earth's biotic, soil, and water resources. The cumulative impacts of land-use change have global consequences, altering the structure and functioning of ecosystems, which in turn can influence the climate system due to the strong linkages between land cover, energy exchange, and biogeochemical cycles. Because of the long timescale dynamics of ecosystem processes, land disturbances can affect ecosystem and climate processes for decades to centuries.

Over geologic timescales, climatic changes associated with changes in Earth's orbit around the Sun have led to large-scale vegetation changes. For example, the Little Ice Age that ended in the 1700s eliminated forests in Iceland and a previously lush green landscape became the now arid region of the Sahara Desert 6,000 years ago (Ritchie et al. 1985). On shorter timescales, severe weather events, fires, herbivory, and human activities have modified Earth's landscapes and converted them to new ecosystems. The impacts of ancient human activities on the landscape have been reviewed extensively (Redman 1999), including the use of fires to maintain open landscapes and the extinction of large Pleistocene mammals after the arrival of humans in North America.

More recently, over the last 300 years, human influence on the land has become globally extensive and intensive (Turner et al. 1990, Foley et al. 2005). Deforestation, agricultural expansion and intensification, desertification, and urban expansion are all significant global environmental issues today (Lepers et al. 2005). Nearly 40 percent of the global land surface is being exploited for agriculture (Foley et al. 2005), and tropical deforestation continues unabated, especially in the Amazon Basin and Southeast Asia (Lepers et al. 2005). Such large-scale changes in land use and land cover can modify regional and global climate, degrade freshwater resources, cause air pollution, fragment habitats, cause species extinction and biodiversity loss, and lead to the emergence of infectious diseases (Foley et al. 2005). Clearly, land-use and land-cover change is a major driver of global change.

Early efforts by geographers and ecologists to compile global vegetation and land-use maps were accomplished through decades of field investigations and consultations and compilation of numerous local, national, and regional vegetation maps, atlases, and other literature (Matthews 1983, Olson et al. 1983, Wilson and Henderson-Sellers 1985). These painstaking efforts took years to achieve, suffered from some degree of subjectivity, and often used sources of varying quality and time periods across different regions. Despite fundamental disagreements in land-cover classes and their distributions (DeFries and Townshend 1994a), they nevertheless greatly improved our understanding of global land-cover and land-use patterns.

The advent of satellite data has revolutionized our ability to characterize global land cover and monitor land-use patterns. Satellite sensors offer a synoptic view of Earth, as well as the objectivity associated with a consistent measurement and methodology for mapping the entire planet. Satellite data have been used to characterize patterns of land-use and land-cover change across the world at scales from a few meters to a few degrees in latitude by longitude depending on the sensor.

In 1972 the National Aeronautics and Space Administration (NASA) launched the Landsat Satellite Program (previously called the Earth Resources Technology Satellite) to study the features of Earth's landscapes and monitor its natural resources (Box 10.1, Figure 10.1). Landsat data demonstrated early success in monitoring Earth's croplands, forests, and other natural resources. It has since become the workhorse for mapping land-use and land-cover change across the world and now provides the longest continuous record of Earth's changing land cover. Moreover, the free availability of epochal global orthorectified Landsat data for the 1990s, 2000s, and so forth, has been a great boon for the land-use and land-cover change community.

BOX 10.1
The Landsat Satellite Program

While weather satellites have been around since the 1960s, there was no systematic remote monitoring of Earth's terrain until the Landsat program (Figure 10.1). Landsat 1 was launched in July 1972 and acquired more than 300,000 images of Earth's land surface using the Multispectral Scanner (MSS) instrument, which recorded data in four spectral bands with 79-m spatial resolution. Seven Landsat missions have been launched since then, with Landsat 7 continuing today. Landsat 1, 2, and 3 missions used the MSS instrument and demonstrated the usefulness of the acquired data for cartography, land surveys, agricultural forecasting, water resource management, forest management, and mapping sea-ice movement. Launched in 1982, Landsat 4 carried the Thematic Mapper (TM) instrument, which is still in wide use today for mapping land-cover change over large areas. The 30-m pixel size combined with seven spectral bands in the visible, near infrared, and midinfrared are well suited for mapping disturbance patterns. The value of Landsat data in land-cover mapping is highlighted by the fact that the current "data gap" in Landsat 7 data due to an instrument malfunction has been a major setback for the scientific community. Landsat 7 is currently not collecting research-grade data, and a follow-up Landsat Data Continuity Mission is therefore being planned.

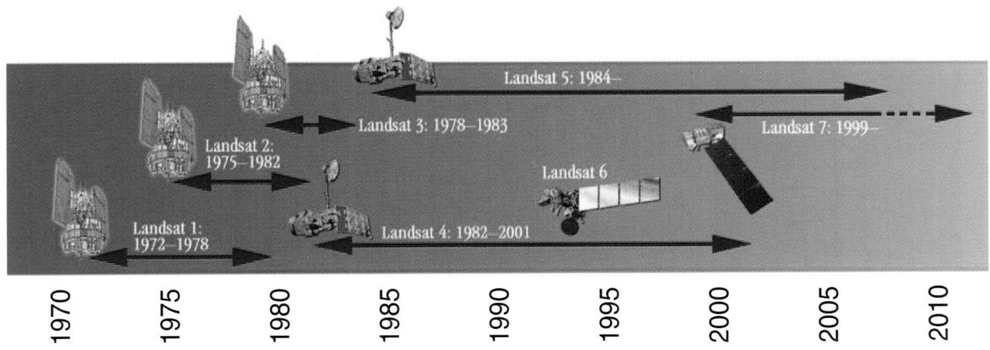

FIGURE 10.1 Timeline of the Landsat satellite series. SOURCE: NASA.

The high cost and effort involved in processing Landsat data over large regions, however, led to the use of coarse- and moderate-resolution sensors (e.g., the Advanced Very High Resolution Radiometer [AVHRR], the Moderate Resolution Imaging Spectroradiometer [MODIS]) during the 1990s and early 2000s. Interestingly, the use of high-resolution commercial data (~1 m; e.g., IKONOS, QUICKBIRD) has become more common recently. Finally, while optical data are best suited for land-cover mapping, active sensors such as radar (e.g., the Japanese Earth Resources Satellite [JERS-1]) are valuable in cloudy regions and also can help derive structural characteristics of vegetation. In summary, technology seems to drive much of the research and applications, but there is always a trade-off in terms of cost and effort involved in processing the data.

MONITORING AGRICULTURAL LANDS

Monitoring food production and forecasting droughts and famines are critical for human societies. Some of the earliest applications of Landsat data included agricultural monitoring and forecasting (Landgrebe 1997). One of the most successful early experiments was LACIE (Large Area Crop Inventory Experiment), begun in November 1974. The capabilities of remote sensing in large-area crop monitoring were demonstrated by LACIE's estimate of wheat production in the Soviet Union during the 1977 growing season to within 6 percent of the reported Soviet figures (MacDonald and Hall 1980). In 1980 this program was broadened to form AgriSTARS (Agriculture and Resources Inventory Surveys Through Aerospace Remote Sensing), which included crop commodity forecasting of all major grains. Similar programs in crop monitoring continue today, such as PECAD

(Production Estimates and Crop Assessment Division) of the Foreign Agricultural Service of the U.S. Department of Agriculture (USDA). The USDA's Cropland Data Layer, developed using Landsat 7 and Advanced Wide Field Sensor (AWiFS) data, is an excellent example of the use of remote sensing to monitor crop patterns and the implications for environment and society (*http://www.nass.usda.gov/research/Cropland/SARS1a.htm*).

Another recent successful application of satellite data in agricultural applications is the Famine Early Warning System Network (FEWS NET). This program was set up in 1985 by the U.S. Agency for International Development, initially in the Sahel and Horn of Africa and now extends to a few other arid developing nations, to incorporate satellite data in famine early warning (Hutchison 1998). This program uses AVHRR data to obtain vegetation conditions and rainfall estimates from the European Meteosat satellite. In FEWS NET, satellite information forms an important component of a multipronged approach to forecasting famines that includes both biophysical information and socioeconomic information to develop indicators for food supply, food access, and levels of development (Hutchison 1998). These and other achievements exemplify the benefits that can be gained from combining satellite observations with other available information (see Box 10.2, Figure 10.2).

ESTIMATING TROPICAL DEFORESTATION

Over the past few decades there has been increasing concern about tropical deforestation and the associated biodiversity loss and environmental consequences. Satellite data have played a crucial role in measuring both the rates and the patterns of forest loss. The first large-scale deforestation map using satellite imagery was made by Tardin and colleagues (1980) for the Brazilian Amazon. The NASA Pathfinder Humid Tropical Deforestation project has since made repeat assessments for the Amazon (Tardin and Cunha 1989, Skole and Tucker 1993) and for much of the tropics (Chomentowski et al. 1994; Figure 10.3).

Deforestation rates have been estimated for the entire tropics in several recent studies. Using a sampling of Landsat scenes, the Food and Agriculture Organization (FAO) mapped tropical deforestation for the 1980s and 1990s (FAO 2001), while the TREES II project of the Joint Research Center of the European Commission mapped deforestation rates for the humid tropics for the 1990s (Achard et al. 2002, 2004). While it is generally acknowledged that high-resolution remote sensing data are needed to identify deforestation, DeFries and colleagues (DeFries et al. 2002, Hansen and DeFries 2004) showed recently that it is also possible to estimate tropical deforestation over large areas using coarse-resolution weather satellite data (8-km resolution AVHRR Pathfinder data) calibrated against high-resolution estimates. Regardless of the specific methods used, all of these satellite-based estimates of deforestation rates were lower than those previously reported by ground-based inventories or national surveys (DeFries and Achard 2002, Hansen and DeFries 2004). The consequence of these new studies has been a lower estimate of carbon emissions from deforestation, with important implications for our understanding of the present-day carbon budget (DeFries and Achard 2002, Houghton 2003, Foley and Ramankutty 2004, Ramankutty et al. 2007).

While satellite data have been widely used to map deforestation around the world, good estimates of selective logging have not been available until recently. Asner and colleagues (2005) developed a method to estimate selective logging over the Amazon Basin using Landsat data (Figure 10.4). The study found that the area of forest damage from selective logging matched or exceeded rates of clear-cut deforestation. This implied a 25 percent increase in the estimate of gross annual anthropogenic emissions of carbon from Amazon forests over that estimated previously from deforestation alone. This has been a remarkable advance in our ability to map fine-scale patterns of land-use practices.

MAPPING GLOBAL LAND COVER

Even though monitoring and identifying regions of rapid land-cover change is a priority for the scientific community (for example, Box 10.3, Figure 10.5), baseline characterization of global land cover and land use is also important, especially for global analysis and modeling of ecosystems and their impacts. As described earlier, it is expensive and laborious to use Landsat data for large-area land-cover mapping. Therefore, moderate-resolution weather satellite sensors (~1-km resolution) have been used to characterize land-cover patterns globally (see Table 10.1). The University of Maryland pioneered the development of global land-cover classification data sets using AVHRR data. Since then there have been at least three other efforts to characterize global land cover (Table 10.1). These efforts have grouped the Earth's landscape into numerous land-cover classes (Figure 10.6). In contrast to the discrete classifiers, the MODIS Vegetation Continuous Fields product provides a continuous description of the landscape (percentage tree cover, herbaceous and bare ground, as well as leaf type and phenology). These global data sets have provided a comprehensive global view of Earth's land surface. They have become valuable inputs for global climate and ecosystem models used to study the influence of land-cover changes on the Earth system (DeFries et al. 1999, Feddema et al. 2005).

MAPPING GLOBAL FIRES

Fires are an important component of ecosystems; many natural communities depend on fires for their regeneration. Natural fires have been around since the presence of oxygen in the atmosphere, and humans have managed fire for more than a half-million years. However, only recently has the

BOX 10.2
Merging Satellite and Ground-Based Data

This chapter mainly discusses approaches to land-cover change research that have directly used remote sensing observations. Many advances, however, have come from approaches that merge satellite data with other ground-based data sources such as census information and survey data. A couple of recent books, *People and Pixels: Linking Remote Sensing and Social Science* (Liverman et al. 1998) and *People and the Environment: Approaches to Linking Household and Community Surveys to Remote Sensing and GIS* (Fox 2003), present several examples of these approaches. Numerous studies have made advances in mapping global land cover, agricultural land-use practices, and urban areas by either merging census and other ancillary information with satellite data using statistical methods or using the ancillary information to guide the land-cover classification from remote sensing (e.g., Ramankutty and Foley 1998, Loveland et al. 2000, Hurtt et al. 2001, Cardille et al. 2002, Frolking et al. 2002, McIver and Friedl 2002, Kerr and Cihlar 2003, Schneider et al. 2003). One example of the "statistical data fusion" approach is the work of Ramankutty et al. (in press), who used global land-cover classification data derived from moderate-resolution remote sensors with national and subnational inventory statistics to develop a global map of the world's croplands (Figure 10.2). Until the advent of remote sensing, our knowledge of the global distribution of agricultural lands was limited to inventory data, which has poor spatial information (available at the administrative level) and is inconsistent in quality across different countries. Therein lies the strength of remote sensing data, which provide consistent and spatially explicit estimates of land-cover across the world. The "data fusion" technique exploits the strengths of both data sources to characterize the world's cultivated lands in a continuous fashion, depicting the percentage of each pixel that is in croplands. The global map indicates that about 12 percent of the global land area is devoted to cultivation and that some areas of the planet are more intensely cultivated than others. This global data set has been useful in various applications such as estimating the carbon cycle and climate implications of land-cover change, estimating global soil erosion, and as providing inputs to global economic models.

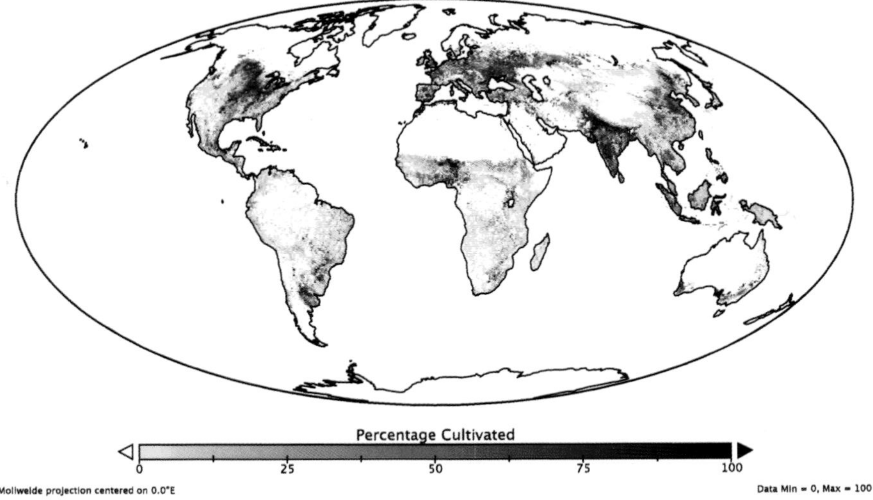

FIGURE 10.2 Croplands of the world in the year 2000. SOURCE: N. Ramankutty.

global distribution of fires been characterized. With the use of remote sensing, rapid progress has been made in documenting the mostly anthropogenic fires in the tropics (Pereira et al. 1999) as well as the primarily natural fires in boreal regions (Kasischke et al. 2002).

Several major efforts have also been undertaken to document fires at the global scale. The Global Burnt Area (GBA-2000) data set derived using the SPOT VEGETATION satellite was the first estimate of the global area of vegetation burned in the year 2000 (Tansey et al. 2004). The ATSR World Fire Atlas (Figure 10.7) is another global inventory of monthly fire maps from 1995 to the present, produced using

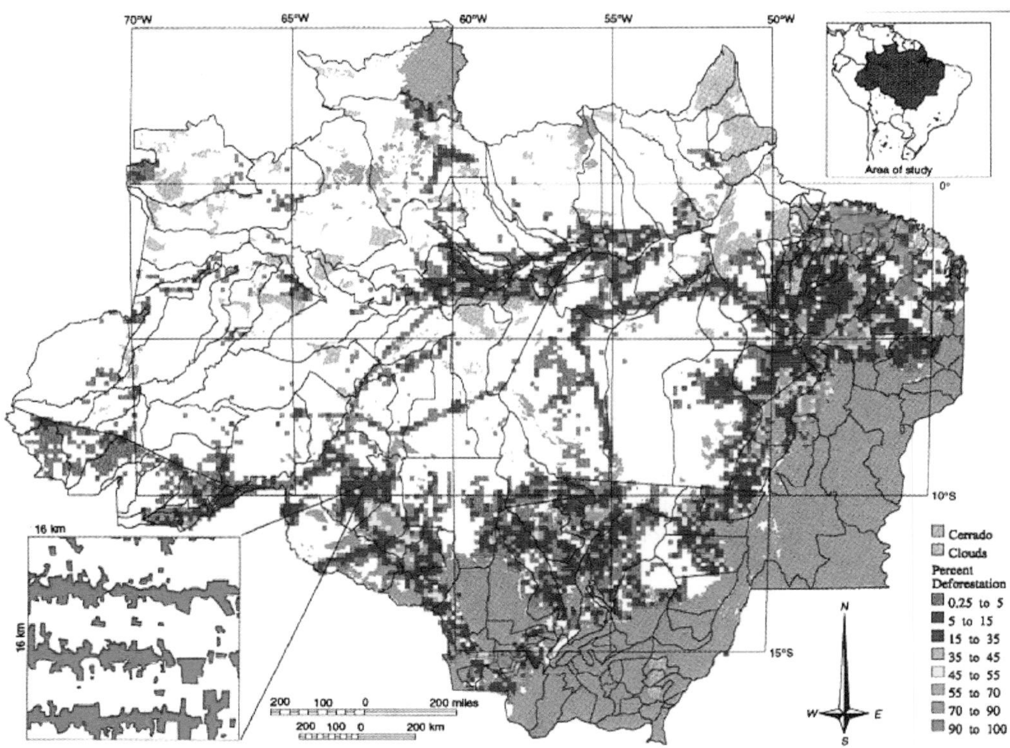

FIGURE 10.3 Quantifying Amazon deforestation in 1988 using NASA Pathfinder Humid Tropical Deforestation project. SOURCE: Skole and Tucker (1993). Reprinted with permission from AAAS, copyright 1993.

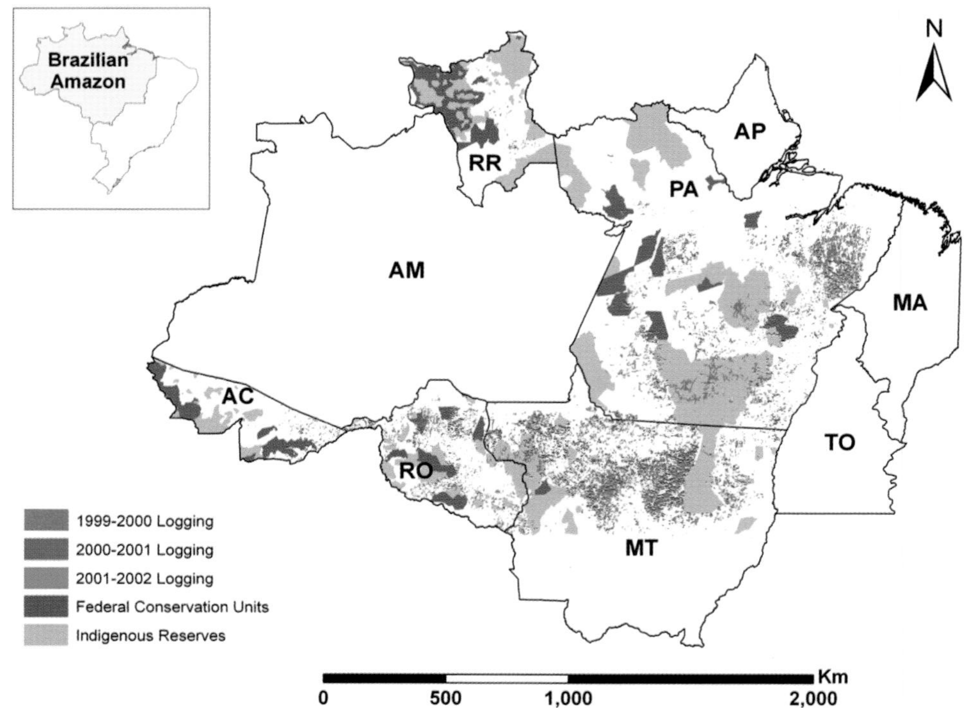

FIGURE 10.4 Estimating selective logging over the Amazon Basin using Landsat data. SOURCE: Asner et al. (2005). Reprinted with permission from AAAS, copyright 2005.

BOX 10.3
Monitoring Urban Areas

Although built-up areas account for less than 2 percent of Earth's land area, more than half of the world's population (3.3 billion people) now live in cities and over 70 percent of economic activity is concentrated in urban areas. Remotely sensed data have played a pivotal role in our ability to monitor, assess, and understand the dynamic processes in urban regions since the early urban land classification efforts of the mid-1970s and following the second generation of satellite sensors (Landsat, SPOT) in the 1980s. The most recent wave of very high resolution sensors and advances in data fusion have spawned new urban remote sensing methods to extract urban features and characterize building materials.

Data from the Landsat sensors have played a particularly important role in assessing urban expansion, primarily because of increased data availability and the synoptic view these data afford. Cities have grown so significantly in the past few decades that it is critical to have accurate and up-to-date maps to help monitor the rate and form of urban and periurban land conversion and to identify how urban expansion differs across cities from a range of geographic settings and levels of economic development. One example of such research is the work of Schneider and Woodcock (in press), who have used a combination of Landsat Thematic Mapper and Enhanced Thematic Mapper data, spatial metrics, and census data to explain differences in urban expansion in a cross-section of 25 midsized cities from around the globe (Figure 10.5). Results show that these patterns can be categorized into a taxonomy of four "city types" as shown in the figure below (yellow indicates the urban extent in 1990; orange shows the increase in urban land from 1990 to 2000). The four city types, or "templates," for growth are *low-growth cities* characterized by modest rates of infilling-type expansion (e.g., Warsaw); *high-growth cities* with rapid, fragmented development (e.g., Bangalore); *expansive-growth cities* with extensive dispersion at low population densities (occurring almost exclusively in U.S. cities, e.g., Washington, D.C.); and *frantic-growth cities*, such as those in China, exhibiting extraordinary rates of growth at high population densities (e.g., Guangzhou). This study also showed that urban patterns outside the United States are not consistent with common conceptions of the American urban sprawl. Although nearly all sample cities are expanding at the urban-rural boundary, results confirm that the majority of non-American cities do not exhibit large, dispersed spatial forms.

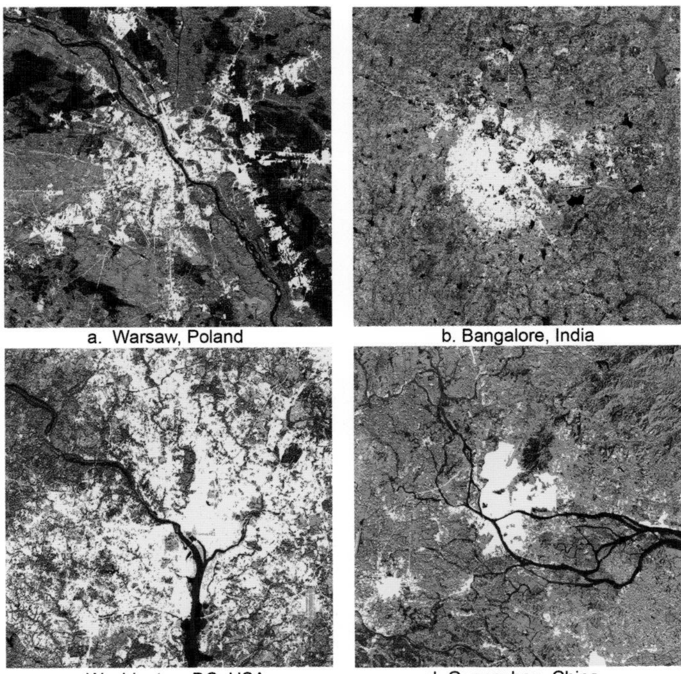

FIGURE 10.5 Urban expansion in four different cities. SOURCE: Schneider and Woodcock (in press). Reprinted with permission from Urban Studies.

TABLE 10.1 Global Land Cover Data Sets from Earth Observation Data.

Data Developer	Name of Product	Sensor	Year of Data	Spatial Resolution	Reference
University of Maryland	UMD Global Land Cover Classification	AVHRR	1987	1 degree	DeFries and Townshend (1994b)
			1984	8-km	DeFries et al. (1998)
			1992	1-km	Hansen et al. (2000)
	Vegetation Continuous Fields MOD44B	MODIS	2001	500-m	Hansen et al. (2003)
U.S. Geological Survey's EROS Data Center; University of Nebraska, Lincoln; and Joint Research Centre, European Commission	Global Land Cover Characterization	AVHRR	1992	1-km	Loveland et al. (2000)
Boston University	MODIS MOD12Q1 Land Cover Product	MODIS	2001	1-km	Friedl et al. (2002)
Joint Research Centre, European Commission	Global Land Cover 2000	SPOT VEGETATION	2000	1-km	Bartholome and Belward (2005)

SOURCE: Ramankutty et al. (2006). Modified by N. Ramankutty, McGill University. Reprinted with kind permission of Springer Science and Business Media, copyright 2006. Modified by Navin Ramankutty, McGill University.

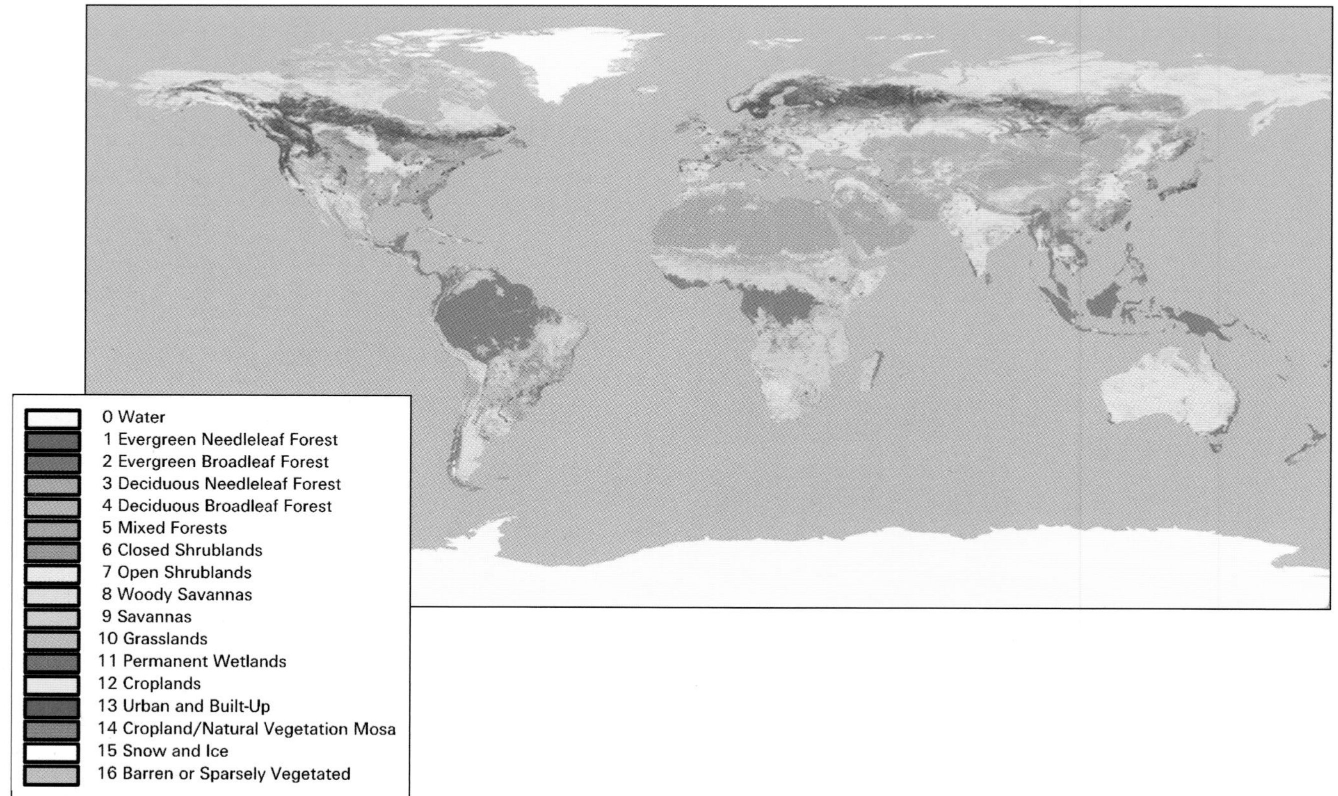

FIGURE 10.6 Earth's land-cover classes. SOURCE: Friedl et al. (2002). Reprinted with permission from Elsevier, copyright 2002.

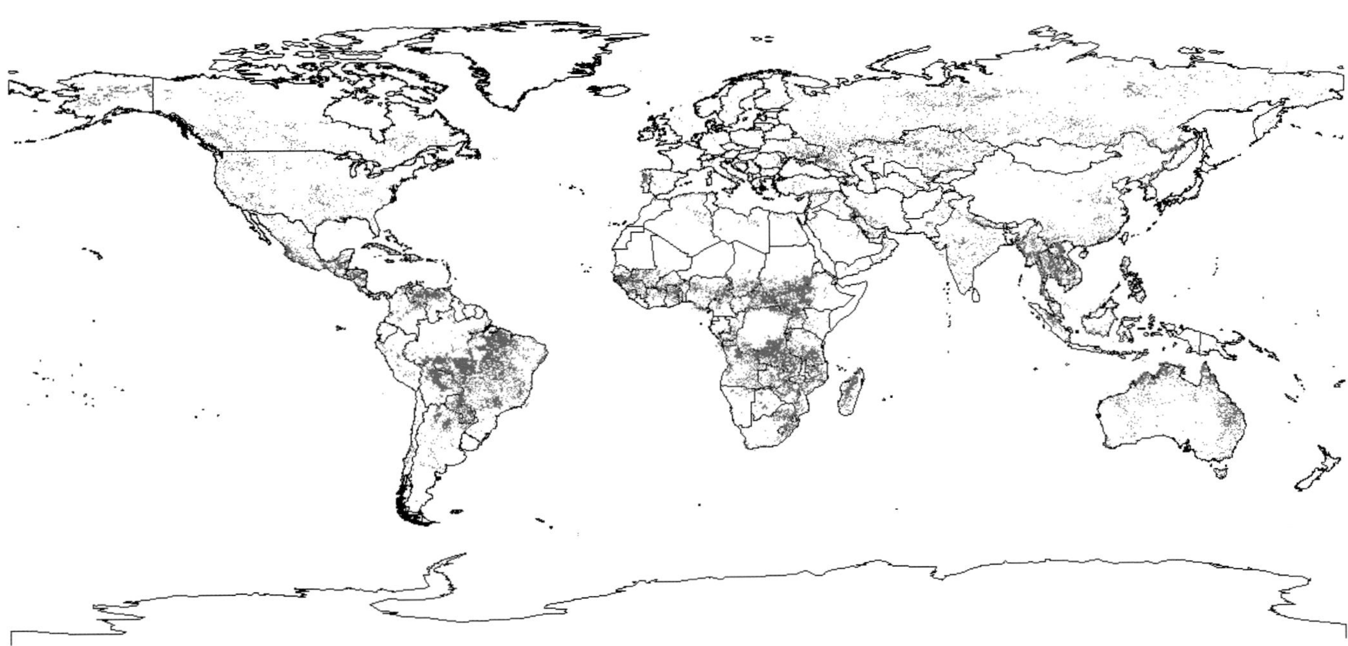

FIGURE 10.7 World Fire Atlas from ATSR. SOURCE: European Space Agency, *http://esamultimedia.esa.int/images/EarthObservation/worldfireatlas_H.jpg*.

the Along Track Scanning Radiometer (ATSR) instrument on the European Remote Sensing (ERS) and ENVISAT satellites (Arino and J.M. Rosaz 1999). GLOBSCAR, a complimentary product to GBA-2000, maps the global distribution of burned area at 1-km spatial resolution and monthly time intervals using the ATSR-2 instrument on the ERS-2 satellite (Simon et al. 2004). These products have been used to compute the emissions of greenhouse gases and aerosols from biomass burning and to explore the impacts on tropical ozone levels (Schultz 2002, Duncan et al. 2003, Palacios-Orueta et al. 2004). Other global fire mapping studies include those of Dwyer et al. (2000), who determined the spatial and seasonal distributions of active fires at the global scale between April 1992 and December 1993, and Riaño et al. (2007), who identified global patterns of fire frequency, seasonality, and periodicity for different land-cover types using 20 years of AVHRR data and established correlations with environmental variables.

UNDERSTANDING DESERTIFICATION

In the 1970s, reports of the southward advance of the Sahara Desert caused increased concern about human-induced desertification (Lamprey 1975, Desert Encroachment Control and Rehabilitation Programme 1976, Smith 1986, Lamprey 1988, Suliman 1988). Based on a survey of about 250 regional soil degradation experts, the Global Assessment of Human-Induced Soil Degradation also reported extensive worldwide desertification (Oldeman et al. 1991). Desertification became the dominant theme of an environmental convention, the United Nations Convention to Combat Desertification, which emerged from the Rio summit of 1992.

Satellite data sets have played a critical role in assessing the role of human activities in desertification. Using the long time series AVHRR record, a study by Tucker et al. (1991) discredited the widely held claims of desertification in the Sahel. The authors found that a satellite-derived vegetation index was highly correlated to measurements of rainfall over the 1980-1990 period, thereby suggesting that vegetation in the Sahel was simply responding to interannual rainfall changes rather than any human-driven causes. Another study by Prince et al. (1998) using AVHRR data for 1982-1990 also found that vegetation productivity was marching in lockstep with precipitation changes and found no evidence for a human hand. Indeed, the wetter conditions prevailing since 1994 seem to be associated with a gradual recovery in vegetation (Anyamba and Tucker 2005). Measuring and attributing desertification remains difficult because a wide variety of environmental changes are taking place at a range of spatial and temporal scales (Reynolds and Stafford-Smith 2002, Reynolds et al. 2007).

11

Solid Earth

Today, solid Earth studies often involve spaceborne techniques, ranging from multispectral imaging to space-geodetic methods. Over the past several decades, space-based observations of the Earth have contributed powerfully to fields such as plate tectonics, seismology, and volcanology, as well as to studies of the geodynamo, mantle convection, and continental tectonics. Such investigations also provide insights into managing natural resources, understanding natural hazards, and predicting global environmental change.

Satellites have revealed Earth's precise shape and how it changes subtly with time and have measured the spatial and temporal changes in mass distribution through measurements of its gravity. Thanks to space geodesy, Earth scientists benefit from an International Earth Reference System that is accurate to better than 1 cm in all components, including the time-dependent position of the geocenter. Even more impressive is the millimeter level relative positioning that is achievable anywhere on the surface of the planet, or in orbit. We can thus measure the movement of tectonic plates in real time and elucidate higher complexities such as the distribution of deformation within plate boundary zones. The transformative nature of this technology is demonstrated by the fact that, a mere 50 years ago, a traveler might not know his/her position on Earth to better than 500 m, even after expending considerable effort in tedious reduction of geodetic observations. Yet, today, an automobilist, aviator, or sailor can determine the vehicle's position to meter precision in real time, anywhere on the planet, using an inexpensive Global Positioning System (GPS) receiver.

GEODESY

National Aeronautics and Space Administration (NASA) satellites have contributed substantially to improving our knowledge of Earth's gravity field. Laser Geodynamics Satellites (LAGEOS) and the Gravity Recovery and Climate Experiment (GRACE) measure Earth's gravity field to model the regional-scale shape of Earth. The shape is irregular and changes over many different timescales (Figures 11.1 and 11.2). The early geoid was described only to the third harmonic degree, revealing the "pear-shaped" departure from the ellipsoid. As more detailed information on Earth's gravity field was made available by LAGEOS and follow-on missions (combined with the expansion to higher harmonic degrees), its precise geoid and topography on a global scale have been made accessible.

Owing to the modern, highly precise, and homogeneous data from satellites such as CHallenging Mission Payload (CHAMP) and GRACE, scientists have been able to derive improved high-resolution global mean gravity field models (Reigber et al. 2003). These models are needed in numerous geodetic-geophysical applications, including the precise orbit determination of Earth satellites, determination of ocean surface currents from altimetry, or GPS leveling. Scientists resolve the gravity anomalies relative to the "idealized" ellipsoidal Earth with the use of these mean gravity models, which have become more sophisticated since the low-orbiting satellites CHAMP and GRACE were able to provide more accurate data. Consequently, these improved gravity models can solve for gravity anomalies 10 times more accurately than before these satellite data became available with direct implications for the aforementioned applications.

STRUCTURE AND DYNAMICS OF EARTH'S DEEP INTERIOR

Satellite measurements of the geoid have provided crucial information to further the understanding of mantle convection. The main features visible in Figures 11.1 and 11.2 emerged in the early global estimates of spatial variations in Earth's gravity field, which incorporated satellite tracking data (e.g., Gaposchkin and Lambeck 1971). Long-wavelength features such as the geoid highs over

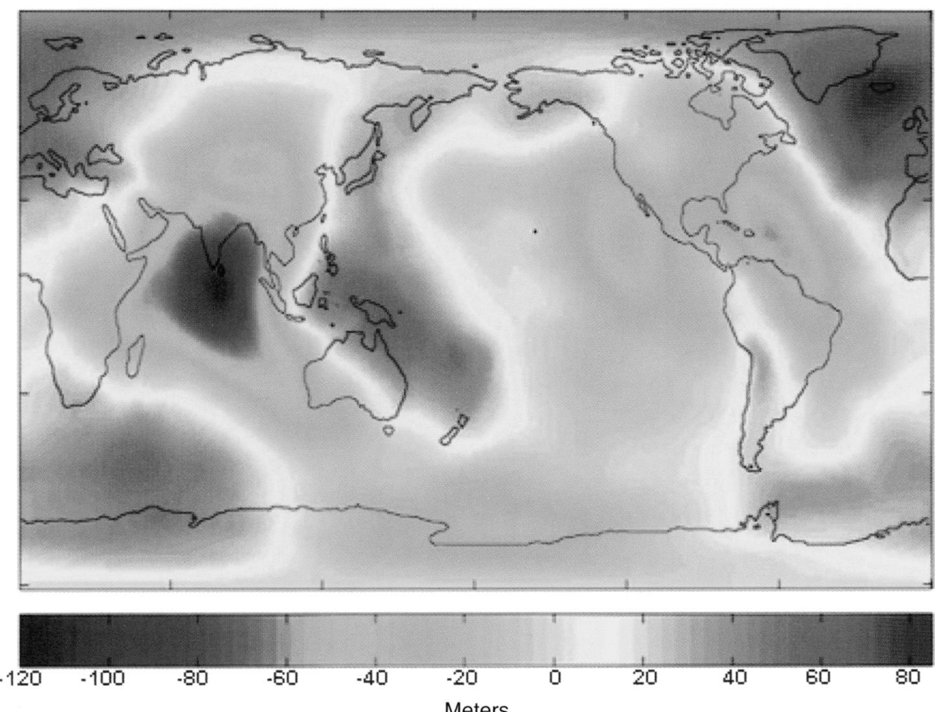

FIGURE 11.1 View of Earth's geoid from the GRACE mission yields deep structures. Using an ellipsoid to approximate the bulk of the Earth's shape and departures from the ellipsoid are represented by the geoid elevation above or below the ellipsoid. The geoid can be as low as 106 m (350 f) below the ellipsoid or as high as 85 m (280 f) above. SOURCE: NASA/Deutsches Zentrum für Luft-und Raumfahrt (DLR).

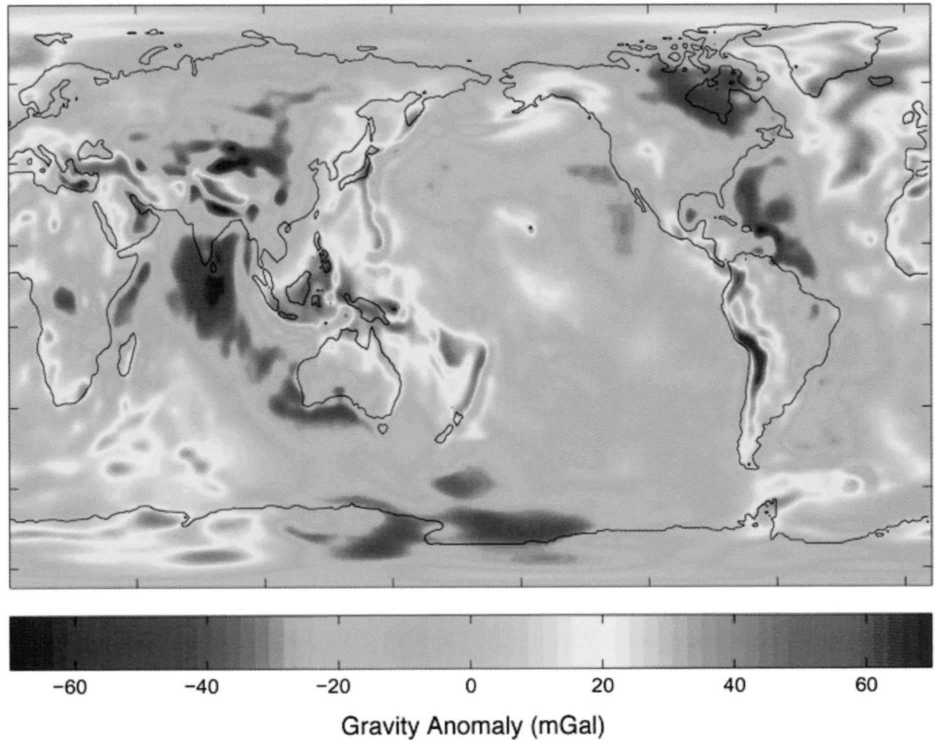

FIGURE 11.2 Earth's gravity anomaly from the GRACE mission yields smaller-scale structures. Standard gravity is defined as the value of gravity for a perfectly smooth "idealized" Earth, and the gravity anomaly (expressed in units of milliGals [mGal]) is a measure of how actual gravity deviates from this standard. SOURCE: NASA/DLR.

regions of plate convergence (New Zealand–New Guinea–Japan–Kamchatka–Aleutians and western South America) indicate mass variations associated with mantle convection and the variation of density and strength in Earth's interior. For example, Hager (1984) demonstrated that these geoid highs over subduction zones require a substantial increase in viscosity with depth between the upper mantle and lower mantle, resulting in an impediment to convective mass transport across the boundary between these two layers. The gravity low over Hudson Bay (Figure 11.2) is due, in part, to a remaining depression in the surface caused by the weight of the great Laurentide ice sheet that melted at the end of the last ice age. The estimate that almost half of this gravity low is the result of ongoing post-glacial rebound again requires a substantial increase in the viscosity of the mantle with depth, otherwise the surface depression would have relaxed more by now (Simons and Hager 1997). The recent observation by GRACE of the rate at which this gravity low is decreasing in amplitude confirms that almost half of this gravity low is the remnant of the former ice sheet (Tamisiea et al. 2007).

THE GLOBAL POSITIONING SYSTEM

NASA missions provided major contributions to the development of the global reference frame through the GPS, Satellite Laser Ranging, and Very Long Baseline Interferometry technology. GPS and Interferometric synthetic aperture radar (InSAR) methods have provided precise measurements of Earth's shape and surface positioning (Box 11.1), thus providing detailed local and global topographic and deformation information. Current InSAR satellites include the European Remote Sensing Satellite (ERS), the European Environmental Satellite (ENVISAT), the Japanese Advanced Land Observation Satellite (ALOS), and the Canadian Radarsat program. These satellites and the constellation of GPS satellites track current motions of Earth's surface at centimeter precision over time and reveal many geophysical processes occurring on the surface and at depth, where they are generally inaccessible to surface observation. Ironically, the use of gravity and deformation data obtained from space has greatly improved our understanding of structure and change deep within the Earth (see below).

BOX 11.1
Earth Reference Frame

Few scientific accomplishments are as "transformative" as the advances in space geodesy over the past five decades, particularly with the ubiquitous introduction of GPS devices. This breakthrough not only has transformed the field of geodesy but also provides vital information for studying global sea-level change, earthquakes, and volcanoes, as well as providing precise position information for all Earth science research.

At the time of the International Geophysical Year, the geolocation of most points at the surface of the Earth entailed errors that reached hundreds of meters in remote areas, even after much effort. Today, scientists rely on an International Earth Reference Frame from which geographical positions can be accurately described relative to the geocenter, in three-dimensional Cartesian coordinates to centimeter accuracy or better—a 2 to 3 orders-of-magnitude improvement compared to 50 years ago. This is true anywhere, on an active planet where every piece of real estate moves relative to every other. Geodesy observations from space have enabled modern measurements of Earth's rotation. The change in position of the rotation axis (the poles) is determined daily to centimeter accuracy, and changes in the length of a day are determined to millisecond accuracy within a few hours. Improvements in GPS measurements over the past few decades have enabled instantaneous geodetic positioning (Genrich and Bock 2006)—a real-time GPS. GPS receivers are now available inexpensively to consumers, who are rapidly becoming accustomed to GPS navigation on the road, on the water, and in the air without realizing the enormous body of science behind this technological achievement: accurate ephemerides of the satellites, very stable clocks, well-calibrated atmospheric corrections, and even relativistic corrections.

PLATE TECTONICS, TOPOGRAPHY, SEISMOLOGY, AND VOLCANOLOGY

The theory of plate tectonics was driven largely by observations in the 1950s from ocean vessels mapping the magnetic field and the seafloor shape, which can now be obtained more easily from satellite observations (Figure 11.3). Several decades later satellite observations enabled a scientific revolution in advancing the theory of plate tectonics by providing highly detailed, quantifiable measurements of Earth's surface. GPS has enabled measurement of plate positioning and velocities, thus resolving geologic and contemporary velocities. For example, Iaffaldano et al. (2006) found that the Nazca Plate moves at a velocity of 6.9 cm per year, compared to its geologic velocity of 10.1 cm per year 10 million years ago. Geologic timescale velocities typically disagree with present rates, with implications for crust-mantle interaction. Factors such as friction or time-dependent processes can be modeled if we understand how the rates vary with time.

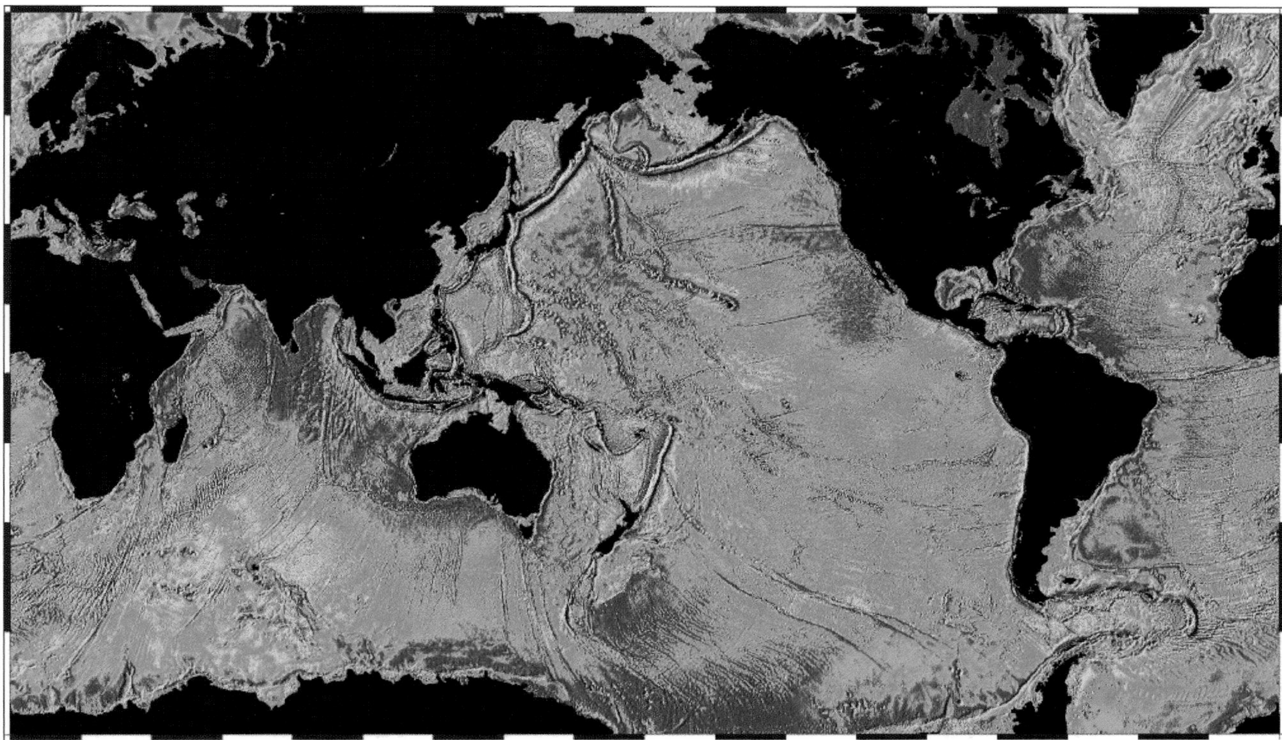

FIGURE 11.3 Map of seafloor topography derived from gravity measurements from satellite tracking and radar altimetry. SOURCE: Reprinted with permission of Dave Sandwell, University of California, San Diego.

Gravity and altimetry measurements from space have also led to discoveries in topography. The Shuttle Radar Topography Mission (SRTM) employed InSAR topography to produce the first (and only) fine-resolution, worldwide, consistent model of elevation. This discovery has mapped the world at 30-m posting, 10-m elevation accuracy; 90-m data are now openly available for Earth. Down-looking radar altimeters measuring ocean heights, which follow the geoid, yield sea-surface topography over the entire ocean at a data density unobtainable on a global scale from shipboard measurements. Applications of detailed gravity information include oil exploration and the location of undersea volcanoes (Smith and Sandwell 2003).

Gravity and topography anomalies relate to large-scale seismic risk and the geophysics of subduction zone boundaries (Song and Simons 2003). Finer-scale risk assessments follow from high-resolution observations of deformation along active faults, which reveal strain accumulations and can indicate stress transfer associated with seismic activity. Therefore, the scientific community took notice after an InSAR observation of the Landers earthquake of 1992, creating the first-ever detailed image of an earthquake and its effect on the crust (Massonnet et al. 1993; Figure 11.4).

Measuring surface displacement is now an important ingredient in seismic risk analysis. For example, Fialko et al. (2002) inferred stress change as a result of the Hector Mine earthquake and the resulting distribution of the stress in the upper crust suggests areas likely for further activity.

Other processes are occurring every day in the solid Earth, many of which escape our knowledge because they occur at a rate slow enough not to radiate seismic energy that can be detected with our present seismographs. Yet these mechanisms for the transfer of energy through the upper crust need to be observed and measured if we are to be able to explain many natural hazards. For example, GPS has enabled the discovery of aseismic ("slow") earthquakes occurring in many subduction zones around the Earth and adding stress to subduction faults (Figure 11.5). The GPS time series for the Cascadia subduction zone shows the result of continual aseismic earthquakes (Melbourne and Webb 2003). Aseismic earthquakes may either dissipate or increase stress, affecting risk probabilities. Unknown until 5 years ago, aseismic earthquakes are a recent discovery dependent on satellite observations.

Inverse methods and the density of InSAR measurements permit a solution for fault slip at depth, giving a view of what is occurring underground as illustrated by images of the Hector Mine earthquake (see Zebker et al. 2000 and Figure 11.6). Such analyses are now commonplace over many terrains.

FIGURE 11.4 Cover of the journal *Nature* showing the first-ever image of an earthquake. This interferogram was produced by combining the pair of ERS-1 SAR images taken before and after the Landers earthquake of June 28, 1992. Each cycle of colored shading represents a range difference of 28 mm between the before and after images, used to detect changes in the position of the ground surface. SOURCE: Massonnet et al. (1993). Reprinted with permission from Macmillan Publishers Ltd., copyright 1993.

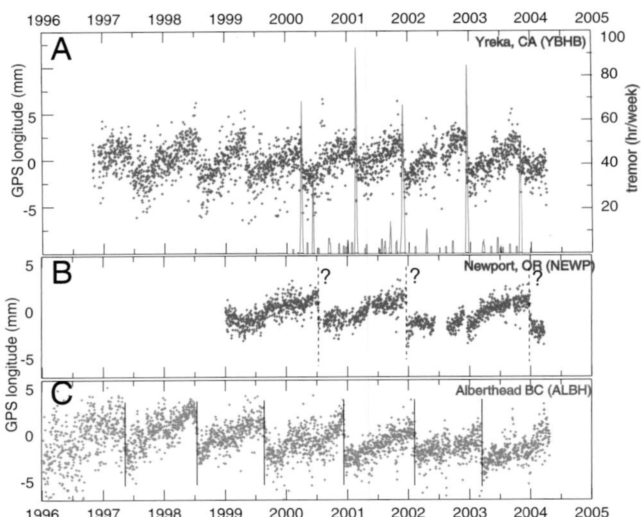

FIGURE 11.5 GPS time series from Yreka, CA (YBHB), Newport, OR (NEWP), and Alberthead, BC (ALBH), and seismic tremor histogram from Yreka. (A) Blue points are daily GPS station positions in mm of the longitudinal component of station YBHB. Solid red line is a plot of the hours of tremor per week at seismic station YBH. Note the similarity of shape displayed by ALBH (C) and YBHB. The correlation between GPS offsets and increased tremor activity indicates that slow faulting occurs beneath Northern California. (B) Purple points represent daily solutions of station position for the longitudinal component of GPS station NEWP from Newport, Oregon. Note the similarity of NEWP offsets (dashed black lines) to those at ALBH. The lack of seismic and continuous GPS stations near NEWP precludes the definitive identification of slow earthquakes here at the present time. (C) Green points represent daily position solutions of the longitudinal component of ALBH. Note the characteristic sawtooth reset shape of the time series due to slow faulting events. Solid black lines denote times of known slow earthquakes at ALBH. SOURCE: Szeliga et al. (2004). Reproduced with permission from American Geophysical Union, copyright 2004.

Many processes on Earth leave strong signals in the deformation of the surface resulting from movements or changes in pressure far beneath the surface. For example, volcanoes cause surface deformation readily observed from satellites. Multiple deformation processes occur simultaneously during a volcanic eruption, prompting the need for volcanic mechanical modeling (Jonsson et al. 2005) in addition to simple mapping of the deformation signature. Detailed observations of the patterns of surficial change allow us to discriminate between many candidate effects and help us better understand the evolution and predictability of volcanoes.

Further applications of spaceborne geodesy follow from measurement of anthropogenic surface change. With implications for natural resource management and natural hazard response, satellites measure subsidence from petroleum extraction (e.g., Lost Hills, CA; Hooper 2005), landslides appearing clearly in InSAR maps (e.g., Berkeley Hills, CA; Hilley et al. 2004), and subsidence from groundwater extraction (Figure 11.7; e.g., Las Vegas Valley, NV, 1992-1997; Amelung et al. 1999).

Water resource managers will be able to model aquifer storage extent (Hoffmann et al. 2003) and begin to map the direction and volume of water migrating through the aquifer system that feeds our cities and farms (see Chapter 6). As described in Chapter 7, gravity measurements have also been applied to studying continental ice sheets. The observed changes in ice flow velocity of glaciers have revolutionized the thinking in how climate change affects the ice sheet mass balance. Measurements of the ice sheet mass balance and maps of ice flow velocity provided by satellites (Rignot 2001) are making important contributions to improved accuracy in forecasting of sea-level rise. These estimates require combining the knowledge gained in solid-Earth geophysics and hydrology, with profound implications for accurately modeling and predicting the consequences of climate change.

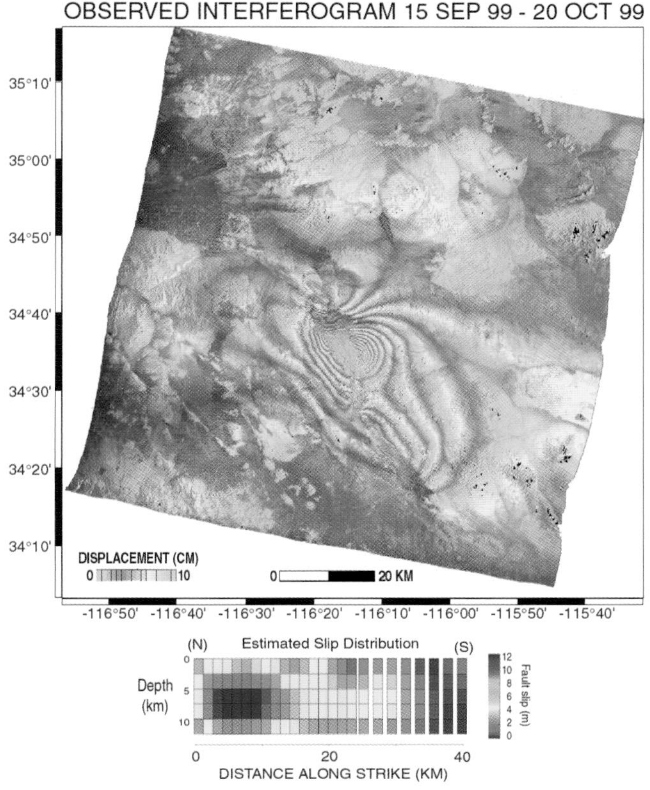

FIGURE 11.6 Hector Mine earthquake, view of fault slip at depth. SOURCE: Zebker et al. (2000). Reproduced with permission from American Geophysical Union, copyright 2000.

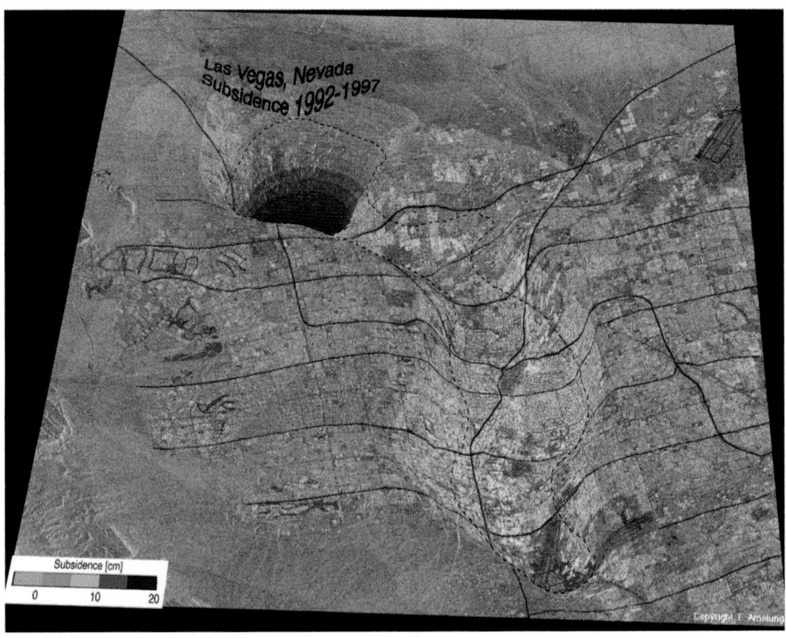

FIGURE 11.7 Subsidence from groundwater extraction in the Las Vegas Valley, 1992-1997. SOURCE: Amelung et al. (1999). Reprinted with permission from the Geological Society of America, copyright 1999.

12

Conclusions

Just as the invention of the mirror allowed humans to see their own image with clarity for the first time, Earth observations from space have allowed humans to see themselves for the first time living on and altering a dynamic planet.

THE EMERGENCE OF INTEGRATED EARTH SYSTEM SCIENCE

During the International Geophysical Year (IGY) of 1957-1958, 67 nations cooperated in an unprecedented effort to study the Earth. In an age otherwise characterized by Cold War tensions, the noted geophysicist Sydney Chapman (1888-1972) referred to the IGY as "the common study of our planet by all nations for the benefit of all." This global effort laid the foundation for the integration of Earth sciences and demanded widespread simultaneous observations. It involved large teams of observers, many of whom were deployed to the ends of Earth—in polar regions, on high mountaintops, and at sea—to study meteorology, oceanography, glaciology, ionospheric physics, aurora and airglow, seismology, gravity, geomagnetism, solar radiation, and cosmic rays. Even in 1957 it was recognized that satellite data would bring observations of Earth that no amount of ground-based observations could achieve.

Hundreds of sounding rockets were launched into the upper atmosphere and near space during the IGY, and the "space age" officially began with geophysical satellites, although still in their infancy, playing an important role (Chapter 2). During the IGY the Soviet Union launched the world's first satellite, Sputnik, in October 1957. The United States launched its first satellite, Explorer 1, shortly thereafter in January 1958. Over the course of the next five decades, the United States and its international partners have launched an array of satellites that fundamentally altered our understanding of the planet. A half-century later, Earth scientists can acquire global satellite data with orders of magnitude greater coverage than obtained during the intensive field expeditions of the IGY from the comfort of their desktops.

The advent of satellites revolutionized the Earth sciences. They provided the first complete global record of biological, physical, and chemical parameters such as cloud cover, winds, and ice cover. They provided consistency of coverage not available with ground measurements. Time series data revealed large-scale processes and features that could not have been discovered by other ways. Prior to the availability of satellite-based observations, scientists seeking global perspectives from largely ground-based observations were required to develop international collaborations and launch large-scale field campaigns. Piecing together data points required interpolation and extrapolation to fill data gaps, particularly for remote locations. In addition, large-scale sampling efforts involved extensive logistics and advance planning, which prohibited frequent repetition. Because the rate of change of many parameters of interest is much greater than the rate at which global maps could be produced in the presatellite era, it was impossible to observe the full dynamics of the system.

Therefore, the unique and revolutionary vantage point from space provides scientists with global images and maps of parameters of interest unmatched by any ground-based observing technology in terms of frequency and coverage. Because satellites collect data continuously and allow for daily (or at least monthly averaged) global images, changes can be observed at the relevant temporal and spatial scale required to detect Earth system processes. The full dynamics of the system have only been observed or characterized since the advent of satellite observations and have allowed the study of previously inaccessible phenomena such as stratospheric ozone creation and depletion, the transport of air pollution across entire ocean basins from China to the continental United States (Chapter 5), global energy fluxes (Chapter 4), ice sheet flow (Chapter 7), global primary productivity (Chapter 9), ocean currents and mesoscale features (Chapter 8), and global maps of winds (Chapter 8). Prior to the satellite era, even if it was possible to compose a global picture from individual surface observations (e.g., through the World Weather Watch, established in 1963), the coverage and density of the network and lack of vertical resolution left much to be desired. Other geophysical and biological phenomena were sampled much less frequently, often as a partial "snapshot" of an otherwise dynamic set of interacting Earth processes.

Discovery of the variability in the velocity of ice sheet flow is another example of how the dynamics of the system went undetected until reliable and repeated satellite observations became available (Chapter 7). This discovery revolutionized the study of ice sheet flow and yielded an important realization: sea-level change due to freshwater input from the continental ice sheets was not a function of the balance between ice sheet melting and precipitation at higher elevation, but a function of the flow dynamics. The increasing velocity of continental ice flow into the ocean in response to climate change and the collapse of the Larsen B Ice Shelf emphasized the sensitivity of ice sheet dynamics to a changing climate.

Satellite sensors provide a panoptic viewpoint, yet historically they suffered from poor resolution and calibration problems. On the other hand, ground-based instruments, although more precise and better calibrated, are limited to their particular locales, and problems arise since they must be coordinated and intercalibrated with other ground stations. As satellite sensors and data processing have become more sophisticated, equaling or surpassing those for ground-based measurements, scientists have obtained not only images but also quantitative global measurements of unprecedented precision. Intercalibration proved particularly challenging in putting together global maps of marine primary productivity from shipboard measurements (Chapter 9). Estimating marine primary productivity requires sample manipulation and measurements of ^{14}C uptake rates at each location, which are sensitive to variations in sampling techniques and methods. Although global marine primary productivity estimates had been attempted before the satellites era, they were flawed because of intercalibration issues. More importantly, because it takes years to obtain global coverage of ground-based marine primary productivity measurements, satellites allowed for the first time observation of global marine primary productivity on a monthly and annual basis and detection of decadal-scale trends.

Satellite observations also provide access to otherwise virtually inaccessible regions, such as polar regions, the upper atmosphere, and the open oceans. Quantitative assessment and monitoring of the sea ice extent in the Arctic has only been possible since routine satellite observations became available. Without satellite images, it is unlikely that trends in decreasing Arctic summer sea ice would have been detected as readily, demonstrating univocally the drastic decline in summer ice over the past decades (Chapter 7). Satellite observations have become available and matured as scientific data at a time when they are critically important in helping society manage planetary-scale resources and environmental challenges. Although many scientific challenges remain, it is undeniable that satellite observations have allowed scientists to improve the ability to monitor and predict changes in the Earth system and manage life on Earth (NRC 2007a).

It is widely known that satellite data, particularly from the southern hemisphere, have contributed to improvements in weather prediction, resulting in protection of human lives and infrastructure (Chapter 3). Since the availability of satellite images, no tropical cyclone has gone undetected, and the advance warning allows crucial time to prepare. In fact, the advent of satellites has been heralded as unquestionably "the greatest single advancement in observing tools for tropical meteorology" (Sheets 1990). Furthermore, because satellite data give access to the largely undersampled ocean, hurricane track forecasts have improved dramatically, helping save lives and property every year (Considine et al. 2004). Other aspects of human welfare have and will also benefit from satellite observations. For example, it is also unlikely that a famine early warning system would be available to assist in planning aid distribution without the ability to observe vegetation cover and the availability of water resources from space (Chapter 10). Given the projected climate change and associated sea-level rise, having global satellite coverage available in the future will serve crucial societal needs unmet by any other observing system.

Conclusion 1: The daily synoptic global view of Earth, uniquely available from satellite observations, has revolutionized Earth studies and ushered in a new era of multidisciplinary Earth sciences, with an emphasis on dynamics at all accessible spatial and temporal scales, even in remote areas. This new capability plays a critically important role in helping society manage planetary-scale resources and environmental challenges.

INTEGRATED GLOBAL VIEW OF THE CARBON CYCLE AND CLIMATE SYSTEM

The global view of Earth from satellites has imparted the understanding that everything is connected—land, ocean, and atmosphere. Interdisciplinary teams of researchers have explored these connections to better understand the Earth as a system beyond the sum of its elements. The concept of studying the Earth as an integrated system at a national level was led by the National Aeronautics and Space Administration (NASA), inspired by NASA's "Ride report" (NASA 1987), and intended as the U.S. component to the International Geosphere-Biosphere Program. Consequently, NASA launched its mission to planet Earth to study the Earth's geosphere and biosphere as an integrated system instead of discrete but interrelated components (CRS 1990).

Other nations have also made significant contributions to the capacity to observe Earth from space. This multinational investment has enabled much international collaboration among satellite projects.

A prime example of an interdisciplinary research endeavor is the study of the global carbon cycle, which employs a wide range of research approaches such as ground and satellite observations, modeling studies, and laboratory experiments. The well-known Keeling curve was obtained from in situ observations and revealed atmosphere-biosphere interactions, as well as the long-term trend of increasing atmospheric carbon dioxide (Keeling et al. 1976). These findings launched major efforts in understanding the role of the terrestrial and oceanic biosphere in carbon uptake through photosynthesis and the impact of increased carbon dioxide levels on global climate. However, primary productivity is controlled by geophysical processes; thus, understanding the interconnections, such as the effect of a changing climate and hydrologic cycle on the global biosphere and vice versa, required observations at a global scale of land-cover changes (from Landsat and AVHRR [Advanced Very High Resolution Radiometer]; see Chapter 11), biomass estimates and primary productivity (AVHRR, CZCS [Coastal Zone Color Scanner], SeaWiFS [Sea-viewing Wide Field-of-view Sensor], and MODIS [Moderate Resolution Imaging Spectroradiometer]; see Chapters 9 and 10), changes in the hydrologic cycle (Landsat, AVHRR, MODIS, and Topography Experiment (TOPEX)/Poseidon; see Chapters 6 and 7) and climate (AVHRR, MODIS, and SeaWiFS). Once the data were available, major scientific advances came from assimilating them into three-dimensional coupled modeling of the atmosphere, land, ocean, and cryosphere (Fung 1986, Heiman and Keeling 1986, Fung et al. 1987, Keeling et al. 1989).

Equally interdisciplinary in nature is climate change research. In fact, many of the accomplishments highlighted in this report have contributed to the improved understanding of the climate system and laid the groundwork modeling for projecting climate change. One notable example is the long-term observations of Earth's radiation budget, which revealed the role of the ocean and atmosphere in transporting heat and the role of aerosols from the volcanic eruption of Mount Pinatubo in cooling the climate (Chapter 4). With the understanding of the importance of aerosols to the climate system comes the need to observe continuously both natural and anthropogenic sources of aerosols (Chapter 4). Satellite observations have also been central in revealing the role of important gases, such as water vapor and ozone, in the climate system (Chapters 4 and 5).

Long-term observations of water in each phase are central to understanding the climate system: sea ice contributes to Earth's albedo and its decrease not only indicates a warmer climate but is also a positive feedback (Chapter 7); melting of continental ice sheets contributes to sea-level rise (Chapter 7); the availability of liquid water is important in controlling the productivity of the terrestrial ecosystem, which in turn affects the amount of carbon dioxide (CO_2) uptake (Chapter 9); and water vapor is important as a greenhouse gas and in heat exchange processes between the ocean, land, and atmosphere (Chapters 3, 4, 8, and 9). Due to water's relatively high specific heat capacity and its large-scale circulation, the ocean plays a central role in storing and transporting Earth's heat content (Chapter 8). In fact, more than 80 percent of Earth's heat is stored in the ocean. Improving our understanding of ocean circulations and consequently the transport of heat is a major challenge to more accurate climate models and predictions. Lastly, the above-mentioned advances in understanding the global carbon cycle further the ability to predict future atmospheric CO_2 levels.

The long-term observations obtained during the past 50 years of Earth science from space combined with advances in data assimilation, computer models, and ground-based process studies brought climate scientists to the point at which they could begin to project how climate change will affect weather and natural resources at the regional level, the scale at which the information is of greatest societal relevance (NRC 2001a).

This comes at a time when improved understanding of the climate system is central to the viability of our economy, as seasonal-to-interannual climate fluctuations strongly influence agriculture, the energy sector, and water resources (CCSP 2003). However, important scientific challenges—for example, cloud-water feedback in climate models—must be conquered with the aid of continuous satellite data before the appropriate seasonal-to-interannual climate information can be made readily available at the appropriate scale (NRC 2007a). The Earth science community has built over the past decades the capacity to incorporate all the pieces into an integrated systems perspective, thanks to ever more sophisticated models. As the community is now poised to make major advances in climate science and predicting climate changes at various scales, the ability to provide sustained multidecadal global measurements is crucial (NRC 1999, 2001b, 2007a).

The ability to observe and predict El Niño/La Niña conditions in advance of their full manifestation based on satellite and in situ data illustrates the significant breakthrough climate scientists have made in providing important regional climate information to resource managers (Box 12.1, Figure 12.1).

As many accomplishments have shown, the length and continuity of a given data record often yield additional scientific benefits beyond the initial research results of the mission and beyond the monitoring implications for operational agencies. For example, the effect of aerosols from a volcanic eruption (Mount Pinatubo) on the global climate would have gone undetected without the continuous observations of the Earth Radiation Budget Experiment (ERBE, Chapter 4). Thus, maintaining well-calibrated long-term data sets is likely to yield important scientific advances in understanding the Earth system, in addition to contributing to societal applications. The importance of stable, accurate, intercalibrated, long-term climate data records is universally recognized, and strategies on how to collect and maintain such data streams have been provided in many previous reports (NRC 1985, 2000, 2001b, 2003, 2004). Important elements to successful long-term climate data from satellites include a long-term strategy to guarantee that follow-on missions overlap to allow for cross-calibrations, leadership in data stewardship and management, and strong interagency collaborations.

Follow-on missions maximize the return on previously made investments in technology development, including sensors and data analysis tools. Missions designed for process studies of initially short durations may provide significant scientific value by continuing a given data record in the context of global change research. The value of a continuous data record increases significantly through the development of uninterrupted follow-on missions, particularly if careful cross-calibrations between subsequent generations of satellite sensors are undertaken (NRC 2004). The long-term data records from Landsat and AVHRR exemplify the scientific value of such carefully maintained data streams (Chapters 9 and 10).

Conclusion 2: To assess global change quantitatively, synoptic data sets with long time series are required. The value of the data increases significantly with seamless and intercalibrated time series (NRC 2004), which highlight the benefits of follow-on missions. Further, as these time series lengthen, historical data sets often increase in scientific and societal value.

MAXIMIZING THE RETURN ON INVESTMENT IN EARTH OBSERVATIONS FROM SPACE

As scientists have gained experience in studying Earth through satellite observations, they have defined new technology needs, helped drive technology development to provide more quantitative and accurate measurements, and advanced more sophisticated methods to interpret satellite data (Chapter 2). Many scientific accomplishments have resulted from rapid satellite technology development that responded to scientific needs and provided capabilities that enabled major advances in the Earth sciences. The value of satellite observations from space grows dramatically as new, more accurate instruments are developed. Initially, satellites provided a means for acquiring pictures. Now, satellite image acquisition and interpretation provide quantitative geophysical or biological variables by transforming measurements of reflected or emitted electromagnetic radiation into desired parameters. For many applications such as ocean and land topography, ice sheet dynamics, and concentrations of atmospheric gases, observations are scientifically valuable if they can be made with great accuracy, which has driven technology evolution. For example, the Williamstown report (NASA 1970) outlines the need for satellite sensors to measure the geopotential and mean sea level to determine the general circulation of the oceans and resolve the spatial variations of the gravity field as a goal for geophysics and physical oceanography. NASA responded to this challenge by launching three satellites within 9 years following the Williamstown conference, with Seasat—the third and most advanced satellite—providing accurate ocean elevation with a precision to tens of centimeters. For the first time the bathymetry of the ocean floor could be observed from space, revealing the large mid-Atlantic ridges and trenches (Chapter 11). As the precision of altimetry data further increased the importance of eddies in the mixing of the open ocean was discovered (Chapter 8).

It is common for any given satellite or instrument in space to supply data that may be used in multiple fields of Earth science by design or serendipitously (see Table A.1). Although Landsat was designed to observe changes on land, including the terrestrial ecosystem, assembling the approximately 5,000 individual images for a global time series proved to be too computationally intensive. Instead, it was AVHRR data—designed to monitor the atmosphere—that turned out to be invaluable to producing global terrestrial primary productivity estimates. Due to careful intercalibrations between the different sensors, the AVHRR data record now extends over 20 years (Chapter 8) and has allowed the detection of trends in terrestrial primary productivity (Chapter 9). In fact, data from AVHRR have also been used in many other fields to study processes such as snow cover, sea surface temperature, cloud optical properties, and global land-cover change (Chapters 6, 8, and 10).

The design of MODIS illustrates the potential for using a single instrument to serve many applications. Its spectral bands were designed to serve a diversity of user communities in the Earth sciences, allowing observations of the following parameters: land, cloud, and aerosol properties; ocean color and marine biogeochemistry; atmospheric water vapor; surface and cloud temperature; cloud properties; cirrus cloud water vapor; atmospheric temperature; ozone; and cloud top altitude. It has led to scientific breakthroughs such as discovery of the brown clouds (Chapter 4), measuring marine primary productivity annually (Chapter 9), and observation of optical depth and effective particle radius in low clouds (Chapter 4). Because of the potential to design missions with spectral bands that can serve many different scientific user communities, creating follow-on missions that continue measurements—and thus ensure the long-term climatic data records discussed above—does not have to come at an increased cost or at the cost of research and development missions.

In addition, the measurement of a given variable, in some cases from multiple sensors, often contributes to several fields of Earth science. For example, few scientific accomplishments are as "transformative" as the advances in space geodesy over the past five decades (Chapter 11). This breakthrough has not only transformed the field of geodesy

BOX 12.1
El Niño-Southern Oscillation

El Niño is a condition that has been known for well over a century. In some years waters off the west coast of South America would become warmer than usual, and the fish populations normally found there would disappear, bringing hardship to fishermen in the region. It occurs periodically around Christmastime and thus was named "El Niño"—the Spanish term referring to the Christ Child.

Much of the groundwork for understanding and describing the El Niño-Southern Oscillation (ENSO) as a coupled atmosphere-ocean phenomenon was laid in the 1970s and 1980s and based on in situ data and modeling studies (e.g., Rowntree 1972, Wyrtki 1975, Rasmusson and Carpenter 1982, Zebiak 1982, Shukla and Wallace 1983, Cane 1984). However, satellite data confirmed observations and model efforts and revealed the global impact of ENSO (Friedler 1984). The improved understanding of the atmosphere-ocean connection has improved the ability to predict ENSO conditions and has advanced our understanding of the teleconnections and impacts on the marine and terrestrial biosphere (Barber and Chavez 1983).

In normal years winds blow from east to west, causing warm surface waters to "pile up" in the western tropical Pacific. During an El Niño, the winds relax and the warm surface waters flow back toward the eastern Pacific. Wind-driven upwellings do not reach deep enough to bring nutrients from below the thermocline. Without the supply of nutrients, phytoplankton do not thrive and this creates a chain reaction in the marine ecosystem. The major El Niño event of 1982 revealed its impacts not only on the ocean but also on global weather patterns, which invigorated research efforts to improve ENSO predictions. Because ENSO events are accompanied typically by drought conditions in Indonesia and Australia and heavier-than-normal rainfall in South America, their effects can be seen in virtually every form of Earth observations from space.

By piecing together the different observations (sea surface temperature [SST], winds, sea surface height, biological productivity, rainfall, and land cover), scientists are working to develop theories to explain what triggers an El Niño and to predict consequences once an El Niño has developed. Satellite observations of SST and winds combined with in situ data are also used to predict El Niño events up to a year in advance. Figure 12.1 illustrates how the physical and biological properties of the Pacific are related during an El Niño and the opposite, La Niña, condition.

FIGURE 12.1 These images of the Pacific Ocean show conditions during an El Niño (1997) and La Niña (1998). The upper images were produced using sea surface height measurements made by the U.S.-French TOPEX/Poseidon satellite. They show variations in sea surface height relative to normal conditions as an indicator of the amount of heat stored in the ocean. The two lower images show variability in chlorophyll concentration relative to normal levels as a measure of phytoplankton biomass. These were produced using data from SeaWiFS. In 1997 the warm surface water in the eastern Pacific (shown in white in the upper figure) was 14 to 32 cm (6 to 13 in.) higher than normal and about 10 cm (4 in.) above normal in the red areas. The same waters were abnormally low in chlorophyll (shown in blue in the lower image) because the supply of nutrients from upwelling was greatly reduced. This El Niño condition results in the well-known absence of fish off the west coast of South America. The images for 1998 show the low sea level or a cold pool of water (shown in purple in the upper image) during the La Niña phase. The lower figure shows higher-than-average chlorophyll (yellow) associated with this cold pool. During La Niña, nutrients were upwelled into the cold pool, resulting in an extensive phytoplankton bloom at the equator that lasted for several months. SOURCE: NASA Jet Propulsion Laboratory (top row); provided by J. Campbell and based on data from SeaWiFS Project, NASA Goddard Space Flight Center, and GeoEye (bottom row).

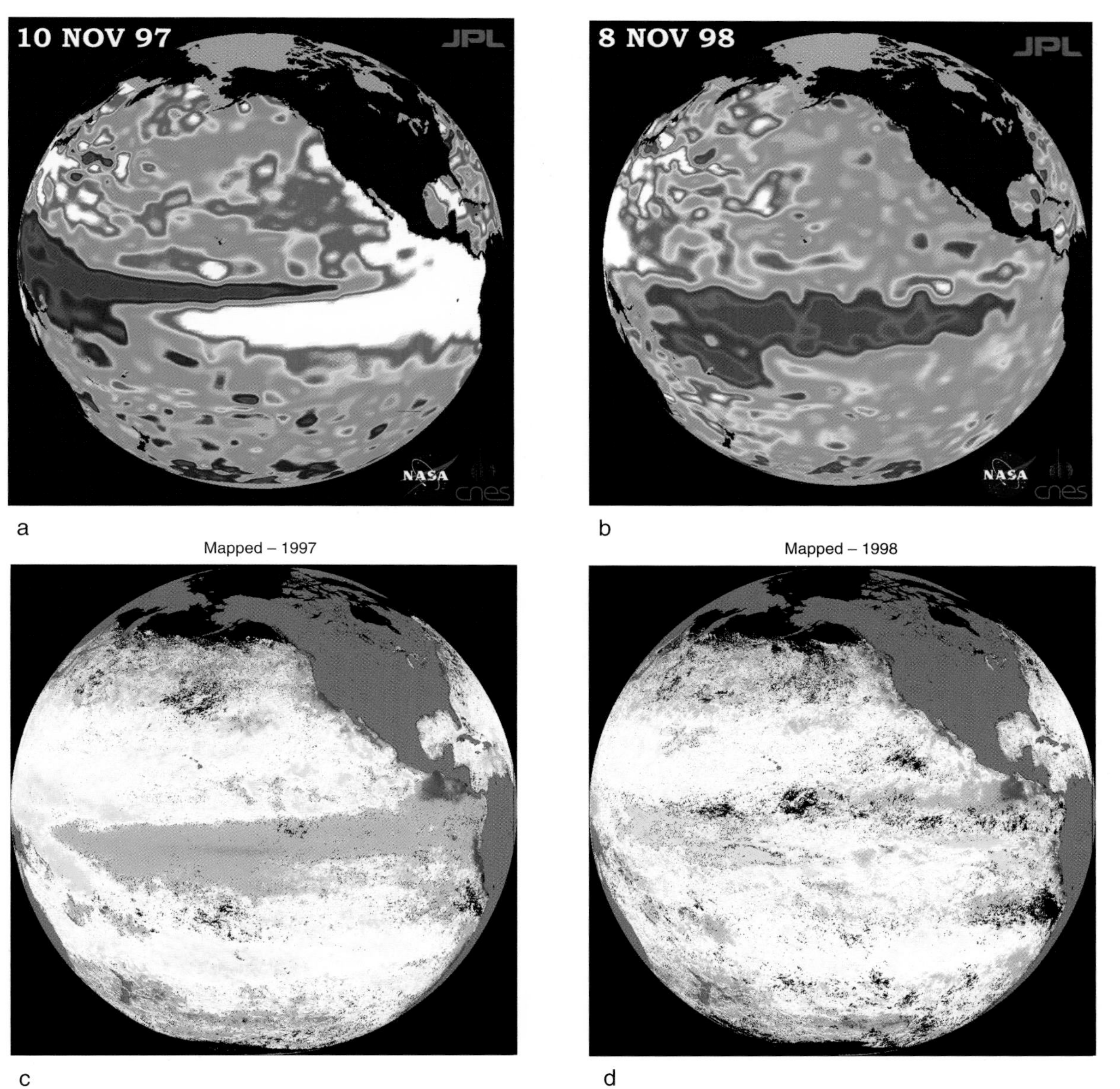
10 NOV 97
JPL
NASA
cnes
8 NOV 98
JPL
NASA
cnes
a
b
Mapped – 1997
Mapped – 1998
c
d

but also provided vital information for studying global sea-level change, earthquakes, and volcanoes. Furthermore, Earth scientists from all disciplines rely on an International Earth Reference Frame from which geographical positions can be accurately described relative to the geocenter, in three-dimensional Cartesian coordinates to centimeter accuracy or better—a 2 to 3 orders of magnitude improvement compared to 50 years ago.

Measured by AVHRR and SAGE (Stratospheric Aerosol and Gas Experiment), aerosols represent a geophysical variable important to Earth's radiation budget, air quality forecasts, cloud formation affecting weather forecasts, and hydrologic applications (Chapter 4). Thus, a scientific accomplishment in one field can lead to major advances in other fields and drive interdisciplinary research efforts. The advances in understanding and predicting El Niño-Southern Oscillation (ENSO) conditions exemplify the advantage of studying the Earth as an integrated system and the benefit of combining in situ and satellite observations with modeling studies.

Conclusion 3: The scientific advances resulting from Earth observations from space illustrate the successful synergy between science and technology. The scientific and commercial value of satellite observations from space and their potential to benefit society often increase dramatically as instruments become more accurate.

The observational vantage point from space added a new appreciation for the complexity of many previously known Earth science processes. Because of the problem of spatial and temporal undersampling by ground-based observing tools, composing a synoptic view required interpolation across data gaps. Consequently, more complex features were averaged out through the interpolation process and not revealed until satellites observed these features directly. Similarly, the frequency of synoptic views available from daily satellite overflights made an unprecedented temporal resolution available. As altimetry measurements became accurate to the centimeter scale, they revealed how highly time dependent and essentially turbulent the ocean was, which is in contrast to the presatellite view that the ocean was primarily in steady state with slowly changing, large-scale circulation (Chapter 8). This resulted in a paradigm shift with implications for climate change research that have yet to be fully understood (Wunsch 2007).

In the case of many scientific accomplishments, significant results are not solely based on satellite data but include in situ data and model components. In fact, the value of space-based observations increases with well-coordinated ground-based observations, suborbital observations, and/or cross-calibration among satellites with complementary instruments. Ground-based observations also provide an important "surface validation" for satellite data and are used to calibrate spaceborne instruments. Such surface validations become increasingly important in pushing satellite sensors to provide more quantitative and accurate measurements. Ocean buoys and drifters as well as shipboard observations have been used extensively to validate sea surface temperature, ocean color, and wind observations from satellites (Chapter 8). In addition, as satellite data have become more quantitative and more readily used by the broader research community, they have contributed to field campaigns and altered the scientific endeavor. For example, ground-based campaigns are more effectively planned and guided because of the information made available from satellite observations.

Just as the synergy between satellite and ground-based observations yields new insights, so does the combination of satellite observations from different instruments. Thus, to capitalize fully on some investments in satellite sensors, simultaneous measurements are necessary. The recent analysis of the merged altimetry data set from TOPEX/Poseidon and the European Remote Sensing Satellite (ERS) revealed the prevalence of westward-propagating eddies not seen from individual sensors (Chapter 8). This discovery would not have been possible without merging the two data sets from the individual sensors.

Conclusion 4: Satellite observations often reveal known phenomena and processes to be more complex than previously understood. This brings to the fore the indisputable benefits of multiple synergistic observations, including orbital, suborbital, and in situ measurements, linked with the best models available.

The greatest benefit of Earth observations from space is gained when data are integrated into state-of-the-art models, combined with ground-based observation network and process studies, and analyzed with sophisticated methods. Model development has aided in developing an interdisciplinary thinking in the Earth sciences. Building sophisticated models and data analysis tools often involves long lead times and requires training of a skilled workforce. Consequently, the major scientific breakthrough might follow years after the satellite data have first become available. To capitalize fully on the investment, satellite data also require careful calibration (NRC 2004). In addition, building long-term data records for climate research requires cross- and intercalibration between various sensors and follow-on missions, data processing and archiving, and maintenance of the metadata (NRC 2004).

To develop the aforementioned infrastructure and data assimilation and analysis tools, scientists need to be trained in using and analyzing satellite data. Thus, investment in training and supporting a remote sensing community is important to guaranteeing scientific advances from satellite data (NRC 2007a). Attracting young scientists to the field of remote sensing is made easier by the prospect of stability in the satellite data supply. In contrast, data gaps may result in the loss of a highly specialized workforce (NRC 2007a). The full benefit of satellite data is only realized when a robust scientific community is trained to use the data to address fundamental and applied research questions.

The Landsat story, described in numerous accounts (e.g., NRC 2002), is a case in point: wholesale commercialization of the data led to a precipitous drop in their use for science and commercial applications, which recovered upon return to the earlier policy that made data access affordable. Only when academic, government, and commercial scientists are given liberal access to data and a sufficient number are trained in the effective use of these data will the analysis tools mature to the benefit of all parties. Similarly, obtaining the maximum benefit from weather satellites required a decade-long process of improving methods of radiance data assimilation (Lord 2006; see Chapter 3).

Conclusion 5: The full benefits of satellite observations of Earth are realized only when the essential infrastructure, such as models, computing facilities, ground networks, and trained personnel, is in place.

NASA's open and free data policy has created a worldwide linked community of Earth scientists. This open-access policy encourages use of the data for scientific purposes and maximizes the potential societal benefits of the observations. The long list of accomplishments is unlikely to have materialized without this open data policy that encouraged the growth of the field (NRC 2004). As previously mentioned, when the Landsat program was privatized during the late 1980s and early 1990s, the data became so costly that it severely hampered the research program (Malakoff 2000), illustrating the importance of maintaining free or affordable data streams.

Open access also increases the societal benefits of the data by allowing nations without the observational capabilities of the developed world to gain access to important environmental observations. The Famine Early Warning System Network, although developed by a U.S. agency, is an example of such an application that aids developing nations in resource management without having to first build the ground-based observational capabilities. Consequently, data sharing among agencies and other countries leads to more than the sum of its parts, particularly if nations with Earth-orbiting satellites collaborate on an international strategy regarding the important satellite missions and data needs to observe the Earth system (NRC 2007a).

Conclusion 6: Providing full and open access to global data to an international audience more fully capitalizes on the investment in satellite technology and creates a more interdisciplinary and integrated Earth science community. International data sharing and collaborations on satellite missions lessen the burden on individual nations to maintain Earth observational capacities.

OPPORTUNITIES FOR THE FUTURE OF EARTH OBSERVATIONS FROM SPACE

Fifty years from now a report similar to this one is likely to describe many more astounding discoveries about the Earth system, if the commitment to satellite observations from space is sustained. Although this report provides an extensive sampling of important accomplishments enabled by Earth satellite data, many scientific questions and societal challenges remain unresolved, including improving 10-day weather forecasts, more accurately forecasting hurricane intensity, increasing resolution of earthquake fault systems and volcanoes to detect precursors of events, mitigating climate change impacts, and protecting natural resources (NRC 2007a).

Because the critical infrastructure to make the best use of satellite data takes decades to build and is now in place, the scientific community is poised to make significant progress toward understanding and predicting the complexity of the Earth system. However, building a predictive capability relies strongly on the availability of seamlessly intercalibrated long-term data records, which can only be maintained if subsequent generations of satellite sensors overlap with their predecessors. Unfortunately, the current capability to observe Earth from space is jeopardized by delays in and lack of funding for many critical satellite missions (NRC 2007a). Because important climate data records and important Earth-observing missions are at risk of suffering detrimental data gaps or of being cut altogether, the committee strongly agrees with the following recommendation by the decadal survey (NRC 2007a):

> **The U.S. government, working in concert with the private sector, academe, the public, and its international partners, should renew its investment in Earth-observing systems and restore its leadership in Earth science and applications.**

To sustain the rate of scientific discovery and advances, committing to the maintenance of long-term observing capacities and to innovation in observing technology is equally important. Because past observations taught scientists that the Earth is a highly dynamic system and not as predictable as initially assumed, long-term observations are required if humans wish to understand and predict future changes. Future advances will be associated with tremendous societal benefits, given the current challenges presented, for example, by climate change and loss of biodiversity. One can envision the availability of regional annual climate predictions to assist in water resource management, an infectious disease early warning system, operational use of air pollution maps, and improved ability to foresee volcanic eruptions or earthquakes (NRC 2001a, 2007a).

The committee strongly agrees with the following lines from the interim report of the decadal survey (NRC 2005):

Understanding the complex, changing planet on which we live, how it supports life, and how human activities affect its ability to do so in the future is one of the greatest intellectual challenges facing humanity. It is also one of the most important challenges for society as it seeks to achieve prosperity, health, and sustainability.

If the nation's commitment to continue Earth observations from space is renewed, we have seen just the beginning of an era of Earth observations from space, and a report in 50 years will be able to highlight many more valuable scientific achievements and discoveries.

Conclusion 7: Over the past 50 years, space observations of the Earth have accelerated the cross-disciplinary integration of analysis, interpretation, and, ultimately, our understanding of the dynamic processes that govern the planet. Given this momentum, the next decades will bring more remarkable discoveries and the capability to predict Earth processes, critical to protect human lives and property. However, the nation's commitment to Earth satellite missions must be renewed to realize the potential of this fertile area of science.

References

Achard, F., H.D. Eva, H.-J. Stibig, P. Mayaux, J. Gallego, T. Richards, and J.-P. Malingreau. 2002. Determination of deforestation rates of the world's humid tropical forests. *Science* 297: 999-1002.

Achard, F., H.D. Eva, P. Mayaux, H.-J. Stibig, and A. Belward. 2004. Improved estimates of net carbon emissions from land cover change in the tropics for the 1990s. *Global Biogeochemical Cycles* 18(2): GB2008 10.1029/2003GB002142.

ACIA. 2005. *Arctic Climate Impact Assessment.* Impacts of a Warming Arctic: Arctic Climate Impact Assessment. Cambridge University Press, Cambridge, UK.

Alsdorf, D.E., E. Rodriguez, and D.P. Lettenmaier. 2007. Measuring surface water from space. *Reviews of Geophysics* 45(RG2002) doi: 10.1029/2006RG000197.

Amelung, F., D.L. Galloway, J.W. Bell, H.A. Zebker, and R.J. Laczniak. 1999. Sensing the ups and downs of Las Vegas; InSAR reveals structural control of land subsidence and aquifer-system deformation. *Geology* 27(6): 483-486.

Anderson, M.C., W.P. Kustas, and J.M. Norman. 2003. Upscaling and downscaling—a regional view of the soil-plant-atmosphere continuum. *Agronomy Journal* 95: 1,408-1,423.

Andrews, D.G., and M.E. McIntyre. 1976. Planetary waves in horizontal and vertical shear: The generalized Eliassen-Palm relation and the mean zonal acceleration. *Journal of Atmospheric Sciences* 33: 2,031-2,048.

Andrews, D.G., J.R. Holton, and C.B. Leovy. 1987. *Middle Atmosphere Dynamics.* Academic Press, New York.

Angert, A., S. Biraud, C. Bonfils, C.C. Henning, W. Buermann, J. Pinzon, C.J. Tucker, and I. Fung. 2005. Drier summers cancel out the CO2 uptake. *Proceedings of the National Academy of Sciences* 102(31) 10,823-10,827.

Antoine, D., J.-M. André, and A. Morel. 1996. Oceanic primary production 2. Estimation at global scale from satellite (coastal zone color scanner) chlorophyll. *Global Biogeochemical Cycles* 10(1): 57-70.

Anyamba, A., C.J. Tucker, and R. Mahoney. 2001. From El Niño to La Niña: Vegetation response patterns over east and southern Africa during the 1997-2000 Period. *Journal of Climate* 15(21): 3,096-3,103.

Anyamba, A., and C. Tucker. 2005. Analysis of Sahelian vegetation dynamics using NOAA-AVHRR NDVI data from 1981-2003. *Journal of Arid Environments* 63: 596-614.

Arino, O., and J.M. Rosaz. 1999. *1997 and 1998 World ATSR Fire Atlas Using ERS-2 ATSR-2 Data.* L.F. Neuenschwander, K.C. Ryan, and G.E. Golberg, eds. University of Idaho and the International Association of Wildland Fire, Boise.

Asner, G.P., and P.M. Vitousek. 2005. Remote analysis of biological invasion and biogeochemical change. *Proceedings of the National Academy of Sciences USA* 102: 4,383-4,386.

Asner, G.P., D.E. Knapp, E.N. Broadbent, P.J.C. Oliveira, M. Keller, and J.N. Silva. 2005. Selective logging in the Brazilian Amazon. *Science* 310: 480-482.

Atkinson, D.E., R. Brown, B. Alt, T. Agnew, J. Bourgeois, M. Burgess, C. Duguay, G. Henry, S. Jeffers, R. Koerner, A.G. Lewkowicz, S. McCourt, H. Melling, M. Sharp, S. Smith, A. Walker, K. Wilson, S. Wolfe, M-K. Woo, and K. Young. 2006. Canadian cryospheric response to an anomalous warm summer. *Atmosphere-Ocean* 44: 347-375.

Bader, M.J., G.S. Forbes, J.R. Grant, R.B.E. Lilley, and A.J. Waters, eds. 1995. *Images in Weather Forecasting: A Practical Guide for Interpreting Satellite and Radar Imagery.* Cambridge University Press, Cambridge, UK.

Barber, R.T., and F.P. Chavez. 1983. Biological consequences of El Niño. *Science* 222: 1,203-1,210.

Barkstrom, B.R. 1984. The Earth Radiation Budget Experiment (ERBE). *American Meteorological Society Bulletin* 65: 1,170-1,185.

Barnett, T.P., D.W. Pierce, and R. Schnur. 2001. Detection of anthropogenic climate change in the world's oceans. *Science* 292: 270-274.

Barnett, T.P., J.C. Adam, and D.P. Lettenmaier. 2005. Potential impacts of a warming climate on water availability in snow-dominated regions. *Nature* 438: 303-309.

Barrett, E.C., and D.W. Martin. 1981. *The Use of Satellite Data in Rainfall Monitoring.* Academic Press, New York.

Barrie, L.A., W. Bottenheim, R.C. Schnell, P.J. Crutzen, and R.A. Rasmussen. 1988. Ozone destruction and photochemical reactions at polar sunrise in the lower Arctic atmosphere. *Nature* 334: 138-141.

Barros, A.P., G. Kim, E. Williams, and S.W. Nesbitt. 2004. Probing orographic controls in the Himalayas during the monsoon using satellite imagery. *Natural Hazards and Earth System Science* 29: 29-51.

Bartholome, E., and A.S. Belward. 2005. GLC2000: A new approach to global land cover mapping from Earth observation data. *International Journal of Remote Sensing* 26: 1,959-1,977.

Behrenfeld, M.J. and P.G. Falkowski. 1997. A consumer's guide to phytoplankton primary productivity models. *Limnology and Oceanography* 42(7): 1,479-1,491.

Behrenfeld, M.J., J.T. Randerson, C.R. McClain, G.C. Feldman, S.O. Los, C.J. Tucker, P.G. Falkowski, C.B. Field, R. Frouin, W.E. Esaias, D.D. Kolber, and N.H. Pollack. 2001. Biospheric primary production during an ENSO transition. *Science* 291: 2,594-2,597.

Behrenfeld, M.J., R.T. O'Malley, D.A. Siegel, C.R. McClain, J.L. Sarmiento, G.C. Feldman, A.J. Milligan, P.G. Falkowski, R.M. Letelier, and E.S. Boss. 2006. Climate-driven trends in contemporary ocean productivity. *Nature* 444: 752-755.

Bellouin, N., O. Boucher, J. Haywood, and M.S. Reddy. 2005. Global estimate of aerosol direct radiative forcing from satellite measurements. *Nature* 438: 1,138-1,141.

Bindschadler, R., P. Vornberger, D.D. Blankenship, T. Scambos, and R. Jacobel. 1996. Surface velocity and mass balance of Ice Streams D and E, West Antarctica. *Journal of Glaciology* 42(142): 461-475.

Bindschadler, R. and P. Vornberger. 1998. Changes in the west Antarctic ice sheet since 1963 from declassified satellite photography. *Science* 279(5,351): 689-692.

Blake, E.S., J.D. Jarrell, E.N. Rappaport, and C.W. Landsea. 2006. The Deadliest, Costliest, and Most Intense United States Tropical Cyclones from 1851 to 2005. NOAA Technical Memorandum NWS TPC-4. Available at http://www.nhc.noaa.gov/Deadliest_Costliest.shtml. Accessed September 24, 2007.

Boccippio, D.J., W. Koshak, R. Blakeslee, K. Driscoll, D. Mach, D. Buechler, W. Boeck, H.J. Christian, and S.J. Goodman. 2000. The Optical Transient Detector (OTD): Instrument characteristics and cross-sensor validation. *Journal of Atmospheric and Oceanic Technology* 17: 441-458.

Brewer, A.W. 1949. Evidence for a world circulation provided by the measurements of helium and water vapor distribution in the stratosphere. *Quarterly Journal of Royal Meteorological Society* 75: 351-363.

Brink, K.H., and T.J. Cowles. 1991. The coastal transition zone program. *Journal of Geophysical Research* 96(C8): 14,637-14,647.

Butler, M.J.A., M.-C. Mouchot, V. Barale, and C. LeBlanc. 1988. The application of remote sensing technology to marine fisheries: An introductory manual. *FAO Fisheries Techchnical Paper* 295.

Cabanes, C., A. Cazenave, and C. Le Provost. 2001. Sea level rise during past 40 years as determined from satellite and in situ observations. *Science* 294: 840-842.

Cane, M.A. 1984. Modeling sea level during El Niño. *Journal of Physical Oceanography* 14: 1,864–1,874.

Cane, M.A., A.C. Clement, A. Kaplan, Y. Kushnir, D. Pozdnyakov, R. Seager, S.E. Zebiak, and R. Murtugudde. 1997. Twentieth-century sea surface temperature trends. *Science* 275: 957-960.

Cardille, J.A., J.A. Foley, and M.H. Costa. 2002. Characterizing patterns of agricultural land use in Amazonia by merging satellite classifications and census data. *Global Biogeochemical Cycles* 16: 1045, doi:10.1029/2000GB001386.

Carn, S.A., L.L. Strow, S. de Souza-Machado, Y. Edmonds, and S. Hannon. 2005. Quantifying tropospheric volcanic emissions with AIRS: The 2002 eruption of Mt. Etna (Italy). *Geophysical Research Letters* 32: L02301, doi:10.1029/2004GL021034.

Carneggie, D.M., S.D. deGloria, and R.N. Colwell. 1974. Usefulness of ERTS-1 and supporting aircraft data for monitoring plant development and range conditions in California's annual grassland. *BLM Final Report* 53500-CT3-266 (N).

Cartwright D.E, and G.A. Alcock. 1983. Altimeter measurements of ocean topography. Pp. 309-319 in *Satellite Microwave Remote Sensing*. T. Allan, ed. Ellis Horwood, Chichester, England.

CCSP (Climate Change Science Program). 2003. (*Our Changing Planet.)* Available at http://www.usgcrp.gov/usgcrp/Library/ocp2003.pdf. Accessed September 27, 2007.

Chahine, M.T. 1968. Determination of the temperature profile of an atmosphere from its outgoing radiance. *Journal of the Optical Society of America* 58: 1,634-1,637.

Chahine, M.T., T.S. Pagano, H.H. Aumann, R. Atlas, C. Barnet, J. Blaisdell, L. Chen, M. Divakarla, E.J. Fetzer, M. Goldberg, C. Gautier, S. Granger, S. Hannon, F.W. Irion, R. Kakar, E. Kalnay, B.H. Lambrigtsen, S. Lee, J. Le Marshall, W.W. McMillan, L. McMillin, E.T. Olsen, H. Revercomb, P. Rosenkranz, W.L. Smith, D. Staelin, L.L. Strow, J. Susskind, D. Tobin, W. Wolf, and L. Zhou. 2006. AIRS: Improving weather forecasting and providing new data on greenhouse gases. *Bulletin of the American Meteorological Society* 87: 911-926.

Chance, K., P.I. Palmer, R.J.D. Spurr, R.V. Martin, T.P. Kurosu, and D.J. Jacob. 2000. Satellite observations of formaldehyde over North America from GOME. *Geophysical Research Letters* 27(21): 3,461-3,464.

Chandra, S., J.R. Ziemke, W. Min, and W.G. Read. 1998. Effects of 1997-1998 El Niño on tropospheric ozone and water vapor. *Geophysical Research Letters* 25: 3,867-3,870.

Chandra, S., J.R. Ziemke, and R.V. Martin. 2003. Tropospheric ozone at tropical and middle latitudes derived from TOMS/MLS residual: Comparison with a global model. *Journal of Geophysical Research* 108(D9): 4291, doi:10.1029/2002JD002912.

Chapin, F.S. III, A.J. Bloom, C.B. Field, and R.H. Waring. 1987. Plant responses to multiple environmental factors. *Bioscience* 37(1): 49-57.

Chapman, S. 1930. Wind mixing and diffusion in the upper atmosphere. *Physical Review* 36: 1,014-1,015.

Chatfield, R.B., H. Guan, A.M. Thompson, and J.C. Witte. 2004. Convective lofting links Indian Ocean air pollution to paradoxical South Atlantic ozone maxima. *Geophysical Research Letters* 31: L06103, doi:10.1029/2003GL018866.

Chatters, G.C., and V.E. Suomi. 1975. The applications of McIDAS. *IEEE Transactions on Geoscience Electronics* GE-13(3): 137-146.

Chelton, D.B., and M.G. Schlax. 1996. Global observations of oceanic Rossby waves. *Science* 272(5259): 234-238.

Chelton, D.B., M.G. Schlax, M.H. Freilich, and R.F. Milliff. 2004. Satellite measurements reveal persistent small-scale features in ocean winds. *Science* 303: 978-983.

Chelton, D.B., M.H. Freilich, J.M. Sienkiewicz, and J.M. Von Ahn. 2006. On the use of QuikSCAT scatterometer measurements of surface winds for marine weather prediction. *Monthly Weather Review* 134: 2,055-2,071.

Chelton, D.B., M.G. Schlax, and R.M. Samelson. 2007a. Summertime coupling between sea surface temperature and wind stress in the California current system. *Journal of Physical Oceanography* 37: 495-517.

Chelton, D.B., M.G. Schlax, R.M. Samelson, and R.A. de Szoeke. 2007b. Global observations of large oceanic eddies. *Geophysical Research Letters* 34: L15606, doi:10.1029/2007GL030812.

Chen, J.L., C.R. Wilson, D.D. Blankenship, and B.D. Tapley. 2006a. Antarctic mass rates from GRACE. *Geophysical Research Letters* 33: L11502, doi:10.1029/2006GL026369.

Chen, J.L., C.R. Wilson, D.D. Blankenship, and B.D. Tapley. 2006b. Satellite gravity measurements confirm accelerated melting of Greenland ice sheet. *Science* 313 (5,795): 1,958-1,960.

Chen, T., W.B. Rossow, and Y. Zhang. 2000. Radiative effects of cloud-type variations. *Journal of Climate* 13: 264-286.

Chomentowski, W., B. Salas, and D.L. Skole. 1994. Landsat Pathfinder project advances deforestation mapping. *GIS World* 7: 34-38.

Christian, H., R. Blakeslee, and S. Goodman. 1989. The detection of lightning from geostationary orbit. *Journal of Geophysical Research* 94: 13,329-13,337.

Chu, D.A., Y.J. Kaufman, C. Ichoku, L.A. Remer, D. Tanre, and B.N. Holben. 2002. Validation of MODIS aerosol optical depth retrieval over land. *Geophysical Research Letters* 29(12): 8007, doi:10.1029/2001GL013205.

Cicerone, R.J., R.S. Stolarski, and S. Walters. 1974. Stratospheric ozone destruction by man-made chlorofluoromethanes. *Science* 185(4,157): 1,165-1,167

Clark, D.K. 1981. Phytoplankton pigment algorithms for the Nimbus-7 CZCS. In *Oceanography from Space*, J.F.R. Gower, ed. Plenum Press, New York.

Cline, D.W., R.C. Bales, and J. Dozier. 1998. Estimating the spatial distribution of snow in mountain basins using remote sensing and energy balance modeling. *Water Resources Research* 34(5): 1,275-1,285.

Coakley, J.A.J., R.L. Bernstein, and P.A. Durkee. 1987. Effect of ship track effluents on cloud reflectivity. *Science* 237: 1020-1022.

Coale, K.H., K.S. Johnson, F.P. Chavez, K.O. Buesseler, R.T. Barber, M.A. Brzezinski, W.P. Cochlan, F.J. Millero, P.G. Falkowski, J.E. Bauer, R.H. Wanninkhof, R.M. Kudela, M.A. Altabet, B.E. Hales, T. Takahashi, M.R. Landry, R.R. Bidigare, X. Wang, Z. Chase, P.G. Strutton, G.E. Friederich, M.Y. Gorbunov, V.P. Lance, A.K. Hilting, M.R. Hiscock, M. Demarest, W.T. Hiscock, K.F. Sullivan, S.J. Tanner, R.M. Gordon, C.N. Hunter, V.A. Elrod, S.E. Fitzwater, J.L. Jones, S. Tozzi, M. Koblizek, A.E. Roberts, J. Herndon, J. Brewster, N. Ladizinsky, G. Smith, D. Cooper, D. Timothy, S.L. Brown, K.E. Selph, C.C. Sheridan, B.S. Twining, and Z.I. Johnson. 2004. Southern Ocean iron enrichment experiment: Carbon cycling in high- and low-Si waters. *Science* 304(5,669): 408-414.

Cohen, J., and D. Entekhabi. 1999. Eurasian snow cover variability and Northern Hemisphere climate predictability. *Geophysical Research Letters* 26(3): 345-348.

Comiso, J.C. 2002. A rapidly declining perennial sea ice cover in the Arctic. *Geophysical Research Letters* 29(20): 1956, doi: 10.1029/2002GL015650.

Connors, V.S., B.B. Gormsen, S. Nolf, and H.G. Reichle, Jr. 1999. Spaceborne observations of the global distribution of carbon monoxide in the middle troposphere during April and October 1994. *Journal of Geophysical Research* 104: 21,455- 21,470.

Conover, J. 1966. Anomalous cloud lines. *Journal of the Atmospheric Sciences* 23: 778-785.

Considine, T.J., C. Jablonowski, B. Posner, and C.H. Bishop. 2004. The value of hurricane forecasts to oil and gas producers in the Gulf of Mexico. *Journal of Applied Meteorology* 43: 1,270-1,281.

Craig, R.A. 1965. *The Upper Atmosphere*. Academic Press, New York.

CRS (Congressional Research Service). 1990. *Mission to Planet Earth and the U.S. Global Change Research Program*. CRS, Washington, D.C.

Crutzen, P.J. 1970. The influence of nitrogen oxides on the atmospheric ozone content. *Quarterly Journal of the Royal Meteorological Society* 96: 320-325.

Darwin, G.H. 1886. On the dynamical theory of the tides of long period. *Proceedings of the Royal Society A* 41: 319-336.

DECARP (Desert Encroachment Control and Rehabilitation Programme). 1976. Sudan's Desert Encroachment Control and Rehabilitation Programme. The General Administration for Natural Resources, Ministry of Agriculture Food & Natural Resources and The Agricultural Research Council, National Council for Research in coll. with UNEP, UNDP and FAO, 227 pp.

Deeter, M.N., L.K. Emmons, G.L. Francis, D.P. Edwards, J.C. Gille, J.X. Warner, B. Khattatov, D. Ziskin, J.-F. Lamarque, S.-P. Ho, V. Yudin, J.-L. Attié, D. Packman, J. Chen, D. Mao, and J.R. Drummond. 2003. Operational carbon monoxide retrieval algorithm and selected results for the MOPITT instrument. *Journal of Geophysical Research—Atmospheres* 108(D14): 4399, doi:10.1029/2002JD003186.

DeFries, R.S., and J.R.G. Townshend. 1994a. Global land cover: Comparison of ground-based data sets to classifications with AVHRR data. Pp. 84-110 in *Environmental Remote Sensing from Regional to Global Scales,* G.F.A.R. Curran, ed. John Wiley, New York.

DeFries, R.S., and J.R.G. Townshend. 1994b. NDVI-derived land cover classification at a global scale. *International Journal of Remote Sensing* 15: 3,567-3,586.

DeFries, R.S., M. Hansen, J.R.G. Townshend, and R. Sohlberg. 1998. Global land cover classifications at 8km spatial resolution: The use of training data derived from Landsat imagery in decision tree classifiers. *International Journal of Remote Sensing* 19: 3141-3168.

DeFries, R.S., C.B. Field, I. Fung, G.J. Collatz, and L. Bounoua. 1999. Combining satellite data and biogeochemical models to estimate global effects of human-induced land cover change on carbon emissions and primary productivity. *Global Biogeochemical Cycles* 13: 803-815.

DeFries, R.S., and F. Achard. 2002. New estimates of tropical deforestation and terrestrial carbon fluxes: Results of two complementary studies. *Land-Use and Land-Cover Change Newsletter* 8: 7-9.

DeFries, R.S., R. A. Houghton, M. C. Hansen, C. B. Field, D. Skole, and J. Townshend. 2002. Carbon emissions from tropical deforestation and regrowth based on satellite observations for the 1980s and 1990s. *Proceedings of the National Academy of Sciences USA* 99: 14,256-14,261.

Demuth, J.L., M. DeMaria, J.A. Knaff, and T.H. Vonder Haar. 2000. An objective method for estimating tropical cyclone intensity and structure from NOAA-15 Advanced Microwave Sounding Unit (AMSU) data. 24th Conference on Hurricanes and Tropical Meteorology, American Meteorological Society, Ft. Lauderdale, 29 May - 2 June 2000.

Demuth, J.L., M. DeMaria, J.A. Knaff, and T.H. Vonder Haar. 2004. Evaluation of advanced microwave sounding unit tropical-cyclone intensity and size estimation algorithms. *Journal of Applied Meteorology* 43: 282-296.

Dessler, A.E., and K. Minschwaner. 2007. An analysis of the regulation of tropical tropospheric water vapor. *Journal of Geophysical Research* 112: D10120, doi:10.1029/2006JD007683.

Diner, D.J., B.H. Braswell, R. Davies, N. Gobron, J.N. Hu, Y.F. Jin, R.A. Kahn, Y. Knyazikhin, N. Loeb, J.P. Muller, A.W. Nolin, B. Pinty, C.B. Schaaf, G. Seiz, and J. Stroeve. 2005. The value of multiangle measurements for retrieving structurally and radiatively consistent properties of clouds, aerosols, and surfaces. *Remote Sensing of the Environment* 97: 495-518.

Dobson, G.M.B. 1956. Origin and distribution of the polyatomic molecules in the atmosphere. *Proceedings of the Royal Society of London A* 236: 187-193.

Dozier, J., and T.H. Painter. 2004. Multispectral and hyperspectral remote sensing of alpine snow properties. *Annual Review of Earth and Planetary Sciences* 32: 465-494.

Dubovik, O., A. Smirnov, B.N. Holben, M.D. King, Y.J. Kaufman, T.F. Eck, and I. Slutsker. 2000. Accuracy assessments of aerosol optical properties retrieved from Aerosol Robotic Network (AERONET) sun and sky radiance measurements. *Journal of Geophysical Research—Atmospheres* 105: 9,791-9,806.

Ducet, N., P.Y. Le Traon, and G. Reverdin. 2000. Global high-resolution mapping of ocean circulation from the combination of T/P and ERS-1/2. *Journal of Geophysical Research* 105: 19,477-19,498.

Duncan, B.N., R.V. Martin, A.C. Staudt, R. Yevich, and J.A. Logan. 2003. Interannual and seasonal variability of biomass burning emissions constrained by satellite observations. *Journal of Geophysical Research—Atmospheres* 108: doi:10.1029/2002JD002378.

Durre, I., R.S. Vose, and D.B. Wuertz. 2006. Overview of the Integrated Global Radiosonde Archive. *Journal of Climate* 19: 53-68.

Dvorak, V.F. 1975. Tropical cyclone intensity analysis and forecasting from satellite imagery. *Monthly Weather Review* 103: 420-430.

Dwyer, E., S. Pinnock, J.-M. Grégoire, and J.M.C. Pereira. 2000. Global spatial and temporal distribution of vegetation fire as determined from satellite observations. *International Journal of Remote Sensing* 21: 1,289-1,302.

Edwards, D.P., J.F. Lamarque, J.L. Attié, L.K. Emmons, A. Richter, J.P. Cammas, J.C. Gille, G.L. Francis, M.N. Deeter, J. Warner, D.C. Ziskin, L.V. Lyjak, J.R. Drummond, and J.P. Burrows. 2003. Tropospheric ozone over the tropical Atlantic: A satellite perspective. *Journal of Geophysical Research* 108(D8): 4,237, doi:10.1029/2002JD002927.

Edwards, D.P., L.K. Emmons, J.C. Gille, A. Chu, J.L. Attie, L. Giglio, S.W. Wood, J. Haywood, M.N. Deeter, S.T. Massie, D.C. Ziskin, and J.R. Drummond. 2006. Satellite-observed pollution from Southern Hemisphere biomass burning. *Journal of Geophysical Research* 111: D14312, doi:10.1029/2005JD006655.

Egbert, G.D., and R.D. Ray. 2000. Significant dissipation of tidal energy in the deep ocean inferred from satellite altimeter data. *Nature* 405: 775-778.

Eisinger, M., and J.P. Burrows. 1998. Tropospheric sulfur dioxide observed by the ERS-2 GOME instrument. *Geophysical Research Letters* 25: 4,177-4,180.

Ellrod, G. 1989. A Decision Tree Approach to Clear Air Turbulence Analysis Using Satellite and Upper Air Data. NOAA Technical Memorandum NESDIS 23. U.S. Department of Commerce, National Oceanic and Atmospheric Administration, Washington D.C.

ESA (European Space Agency). 1993. *ERS -1 User Handbook*, Paris.

FAO (Food and Agriculture Organization). 2001. FRA 2000: Pan-Tropical Survey of Forest Cover Changes 1980-2000. FRA Working Paper No. 49. FAO, Rome.

Farman, J.C., B.G. Gardiner, and J.D. Shanklin. 1985. Large losses of total ozone in Antarctica reveal seasonal ClO_x/NO_x interaction. *Nature* 315: 207-210.

Farr, T.G., P.A. Rosen, E. Caro, R. Crippen, R. Duren, S. Hensley, M. Kobrick, M. Paller, E. Rodriguez, L. Roth, D. Seal, S. Shaffer, J. Shimada, J. Umland, M. Werner, M. Oskin, D. Burbank, and D. Alsdorf. 2007. The Shuttle Radar Topography Mission. *Reviews of Geophysics* 45: RG2004, doi:10.1029/2005RG000183.

Feddema, J.J., K.W. Oleson, G.B. Bonan, L.O. Mearns, L.E. Buja, G.A. Meehl, and W.M. Washington. 2005. The importance of land-cover change in simulating future climates. *Science* 310: 1,674-1,678.

Fekete, B.M., C.J. Vörösmarty, J. Roads, and C. Willmott. 2004. Uncertainties in precipitation and their impacts on runoff estimates. *Journal of Climate* 17: 294-304.

Fialko, Y., D. Sandwell, D. Agnew, M. Simons, P. Shearer, and B. Minster. 2002. Deformation on nearby faults induced by the 1999 Hector Mine earthquake. *Science* 297(5,588): 1,858-1,862.

Fiedler, P.C., and R.M. Laurs. 1990. Variability of the Columbia River plume observed in visible and infrared satellite imagery. *International Journal of Remote Sensing* 11: 999–1010.

Field, C.B., J.T. Randerson, and C.M. Malmstrom. 1995. Global net primary production: Combining ecology and remote sensing. *Remote Sensing of Environment* 51: 74-88.

Field, C.B., M.J. Behrenfeld, J.T. Randerson, and P. Falkowski. 1998. Primary production of the biosphere: Integrating terrestrial and oceanic components. *Science* 281: 237-240.

Fishman, J., and J.C. Larsen. 1987. Distribution of total ozone and stratospheric ozone in the tropics: Implications for the distribution of tropospheric ozone. *Journal of Geophysical Research* 92: 6,627-6,634.

Fishman, J., K. Fakhruzzaman, B. Cros, and D. Nganga. 1991. Identification of widespread pollution in the Southern Hemisphere deduced from satellite analyses. *Science* 252: 1,963-1,696.

Fishman, J., A.M. Thompson, R.D. Diab, G.E. Bodeker, M. Zunckel, G.J.R. Coetzee, C.B. Archer, D.P. McNamara, K.E. Pickering, J. Combrink, J. Fishman, and D. Nganga. 1996. Ozone over southern Africa during SAFARI-92/TRACE A. *Journal of Geophysical Research* 101(D19): 23,793-23,808.

Fishman, J., A.E. Wozniak, and J.K. Creilson. 2003. Global distribution of tropospheric ozone from satellite measurements using the empirically corrected tropospheric ozone residual technique: Identification of the regional aspects of air pollution. *Atmospheric Chemistry and Physics Journal* 3: 893-907.

Fishman, J., J.K. Creilson, A.E. Wozniak, and P.J. Crutzen. 2005. Interannual variability of stratospheric and tropospheric ozone determined from satellite measurements, *Journal of Geophysical Research* 110: D20306, doi:10.1029/2005JD005868.

Flanner, M.G., and C.S. Zender. 2006. Linking snowpack microphysics and albedo evolution. *Journal of Geophysical Research* 111: D12208, doi:10.1029/2005JD006834.

Foley, J.A., and N. Ramankutty. 2004. A primer on the terrestrial carbon cycle: What we don't know but should. Pp. 279-294 in *The Global Carbon Cycle: Integrating Humans, Climate, and the Natural World*, vol. 62, C.B. Field and M.R. Raupach, eds. Island Press, Washington, D.C.

Foley, J.A., R. DeFries, G.P. Asner, C. Barford, G. Bonan, S.R. Carpenter, F.S. Chapin, M.T. Coe, G.C. Daily, H.K. Gibbs, J.H. Helkowski, T. Holloway, E.A. Howard, C.J. Kucharik, C. Monfreda, J.A. Patz, I.C. Prentice, N. Ramankutty, and P.K. Snyder. 2005. Global consequences of land use. *Science* 309: 570-574.

Folkins, I., and R.V. Martin. 2005. The vertical structure of tropical convection, and its impact on the budgets of water vapor and ozone. *Journal of Atmospheric Science* 62: 1,560-1,573.

Fox, J. 2003. *People and the Environment: Approaches for Linking Household and Community Surveys to Remote Sensing and GIS*. Kluwer Academic Publishers, Boston.

Frei, A., and D.A. Robinson. 1999. Northern Hemisphere snow extent: Regional variability 1972-1994. *International Journal of Climatology* 19(14): 1,535-1,560.

Friedl, M.A., D.K. McIver, J.C.F. Hodges, X.Y. Zhang, D. Muchoney, A.H. Strahler, C.E. Woodcock, S. Gopal, A. Schneider, A. Cooper, A. Baccini, F. Gao, and C. Schaaf. 2002. Global land cover mapping from MODIS: Algorithms and early results. *Remote Sensing of Environment* 83: 287-302.

Friedler, P.C. 1984. Satellite observations of the 1982-1983 El Niño along the US Pacific Coast. *Science* 224: 1,251-1,254.

Friedman, L.D. 2006. Remembering James Van Allen. Planetary News. Available at http://www.planetary.org/news/2006/0810_Remembering_James_Van_Allen.html. Accessed April 20, 2007.

Frohlich, C., and J. Lean. 2004. Solar radiative output and its variability: Evidence and mechanisms. *Astronomy and Astrophysics Review* 12: 273-320.

Frolking, S., J.J. Qiu, S. Boles, X.M. Xiao, J.Y. Liu, Y.H. Zhuang, C.S. Li, and X.G. Qin. 2002. Combining remote sensing and ground census data to develop new maps of the distribution of rice agriculture in China. *Global Geochemical Cycle* 6(4): 1091, doi:10.1029/2001GB001425.

Fu, L.L., and D. Chelton. 2001. Large scale ocean circulation. In *Satellite Altimetry and Earth Sciences*, L.L. Fu and A. Cazenave, eds. Academic Press, New York.

Fuglister, F.C., and L.V. Worthington. 1951. Some results of a multiple ship survey of the Gulf Stream. *Tellus* 3: 1-14.

Fujita, T.T. 1978. Manual of downburst identification for project NIMROD. Research Paper No. 156. University of Chicago, Department of Geophysical Sciences.

Fung, I. 1986. Analysis of the seasonal and geographic patterns of atmospheric CO_2 distributions with a 3-D tracer model. In *The Changing Carbon Cycle: A Global Analysis*, J.R. Trabalka and D.E. Reichle, eds. Springer-Verlag, New York.

Fung, I.Y., C.J. Tucker, and K.C. Prentice. 1987. Application of advanced very high-resolution radiometer vegetation index to study atmosphere-biosphere exchange of CO_2. *Journal of Geophysical Research—Atmospheres* 92(D3): 2,999-3,015.

Gamon, J.A., J. Peñuelas, and C.B. Field. 1992. A narrow-waveband spectral index that tracks diurnal changes in photosynthetic efficiency. *Remote Sensing of Environment* 41(1): 35-44.

Gao, B.-C., and A.F.H. Goetz. 1995. Retrieval of equivalent water thickness and information related to biochemical components of vegetation canopies from AVIRIS data. *Remote Sensing of Environment* 52: 155-162.

Gaposchkin, E.M., and K. Lambeck. 1971. Earth's gravity field to the sixteenth degree and station coordinates from satellite and terrestrial data (Earth gravity field and satellite tracking stations positions geodetic parameters in geocentric reference frame). *Journal of Geophysical Research* 76: 4,855-4,883.

Garrett, C. 2003. Internal tides and ocean mixing. *Science* 301: 1,858-1,859.

Genrich, J.F., and Y. Bock. 2006. Instantaneous geodetic positioning with 10-50 Hz GPS measurements: Noise characteristics and implications for monitoring networks. *Journal of Geophysical Research* 111: B03403.

Gettelman, A., W.J. Randel, F. Wu, and S.T. Massie. 2002. Transport of water vapor in the tropical tropopause layer. *Geophysical Research Letters* 29(1): 1009, doi:10.1029/2001GL013818.

Gille, J.C., and F.B. House. 1971. On the inversion of limb radiance measurements I: Temperature and thickness. *Journal of the Atmospheric Sciences* 28(8): 1,427-1,442

Goody, R.M. 1954. *The Physics of the Stratosphere*. Cambridge University Press, New York City.

Goody, R. 1982. *Global Change: Impacts on Habitability*. Report by the executive committee of a workshop held at Woods Hole, Massachusetts, June 16-21, 1982. JPL D-95. National Aeronautics and Space Administration, Jet Propulsion Laboratory, California Institute of Technology, Pasadena.

Gordon, H.R., D.K. Clark, J.L. Mueller, and W.A. Hovis. 1980. Phytoplankton pigments from the Nimbus-7 coastal zone color scanner: Comparisons with surface measurements. *Science* 210: 63-66.

Gordon, H.R., and A. Morel. 1983. *Remote Assessment of Ocean Color for Interpretation of Satellite Visible Imagery*. Springer-Verlag, New York.

Hager, B.H. 1984. Subducted slabs and the geoid: Constraints on mantle rheology and flow. *Journal of Geophysical Research* 89: 6,003-6,016.

Hagfors, T., and K. Schlegel. 2001. Earth's ionosphere. In *The Century of Space Science*, J.A.M. Bleeker, J. Geiss, and M.H. Dordrecht, eds. Kluwer Academic Publishers, New York.

Hall, D.K., G.A. Riggs, V.V. Salomonson, N. DiGiromamo, and K.J. Bayr. 2002. MODIS snow-cover products. *Remote Sensing of Environment* 83(1-2): 181-194.

Hansen, J., A. Lacis, R. Ruedy, and M. Sato. 1992. Potential climate impact of the Mount Pinatubo eruption. *Geophysical Research Letters* 19: 215-218.

Hansen, J., L. Nazarenko, R. Ruedy, M. Sato, J. Willis, A. Del Genio, D. Koch, A. Lacis, K. Lo, S. Menon, T. Novakov, J. Perlwitz, G. Russell, G.A. Schmidt, and N. Tausnev. 2005. Earth's energy imbalance: Confirmation and implications. *Science* 308 (5,727): 1,431-1,435.

Hansen, M.C., R.S. DeFries, J.R.G. Townshend, and R. Sohlberg. 2000. Global land cover classification at 1 km spatial resolution using a decision tree classifier. *International Journal of Remote Sensing* 21: 1,331-1,365.

Hansen, M.C., and R.S. DeFries. 2004. Detecting long-term global forest change using continuous fields of tree-cover maps from 8-km Advanced Very High Resolution Radiometer (AVHRR) data for the years 1982-1999. *Ecosystems* 7: 695-717.

Harris, G.W., F.G. Wienhold, and T. Zenker. 1996. Airborne observations of strong biogenic NO_x emissions from the Namibian Savanna at the end of the dry season. *Journal of Geophysical Research* 101(D19): 23,707-23,712.

Harrison, E.F., P. Minnis, B.R. Barkstrom, V. Ramanathan, R.D. Cess, and G.G. Gibson. 1990. Seasonal variation of cloud radiative forcing derived from the Earth Radiation Budget Experiment. *Journal of Geophysical Research* 95: 18,687-18,703.

Hartmann, D.L., M.E. Ockert-Bell, and M.L. Michelsen. 1992. The effect of cloud type on Earth's energy—balance—global analysis. *Journal of Climate* 5: 1,281-1,304.

Heiman, M., and C.D. Keeling. 1986. Meridional eddy diffusing model of the transport of atmospheric carbon dioxide. 1. The seasonal carbon cycle over the tropical Pacific Ocean. *Journal of Geophysical Research* 91: 7,765-7,781.

Herman, J.R., P.K. Bhartia, O. Torres, C. Hsu, C. Seftor, and E. Celarier. 1997. Global distribution of UV-absorbing aerosols from Nimbus 7/TOMS data. *Journal of Geophysical Research—Atmospheres* 102: 16,911-16,922.

Hickey, J.R., F. House, H. Jacobowitz, R.H. Maschhoff, P. Pellegrino, L.L. Stowe, and T.H. Vonder Haar. 1980. Initial solar irradiance determination from Nimbus 7 cavity radiometer measurements. *Science (Reports)* 208: 281-283.

Hilley, G.E., R. Bürgmann, A. Ferretti, F. Novali, and F. Rocca. 2004. Dynamics of slow-moving landslides from permanent scatterer analysis. *Science* 304(5679):1952-1955.

Hirota, I., and J.J. Barnett. 1977. Planetary waves in the winter mesosphere—preliminary analysis of Nimbus 6 PMR results. *Quarterly Journal of the Royal Meteorological Society* 103: 487-498.

Hoffmann, J., D.L. Galloway, and H.A. Zebker. 2003. Inverse modeling of interbed storage parameters using land subsidence observations, Antelope Valley, California. *Water Resources Research* 39(2): 1,031, doi:10.1029/2001WR001252.

Holben, B.N., T.F. Eck, I. Slutsker, D. Tanre, J.P. Buis, A. Setzer, E. Vermote, J.A. Reagan, Y.J. Kaufman, T. Nakajima, F. Lavenu, I. Jankowiak, and A. Smirnov. 1998. AERONET—A federated instrument network and data archive for aerosol characterization. *Remote Sensing of Environment* 66: 1-16.

Holland, M.M., and C.M. Bitz. 2003. Polar amplification of climate change in coupled models. *Climate Dynamics* 21(3-4): 221-232.

Hollwedel, J., M. Wenig, S. Beirle, S. Kraus, S. Kühl, W. Wilms-Grasse, U. Platt, and T. Wagner. 2004. Year-to-year variations of spring time polar tropospheric BrO as seen by GOME. *Advances in Space Research* 34: 304-308.

Hooper, B., L. Murray, and C. Gibson-Poole, (eds.). 2005. The Latrobe Valley CO_2 Storage Assessment. Cooperative Research Centre for Greenhouse Gas Technologies, Canberra. CO2CRC Publication No RPT05-0220.

Houghton, J.T. 1984. *Remote Sounding of Atmospheres*, J.T. Houghton, F.W. Taylor, and C.D. Rodgers, eds. Cambridge University Press, New York.

Houghton, R.A. 2003. Why are estimates of the terrestrial carbon balance so different? *Global Change Biology* 9: 500-509.

House, F.B., A. Gruber, G.E. Hunt, and A.T. Mecherikunnel. 1986. History of satellite missions and measurements of the Earth radiation budget (1957-1984). *Review of Geophysics* 24: 357-377.

Hubert, L.F., and O. Berg. 1955. A rocket portrait of a tropical storm. *Monthly Weather Review* 83(6): 119-124.

Hubert, L.F., and L.F. Whitney, Jr. 1971. Wind estimation from geostationary-satellite pictures. *Monthly Weather Review* 99: 665-672.

Hunt, B.G., and S. Manabe. 1968. An investigation of thermal tidal oscillations in the Earth's atmosphere using a general circulation model. *Monthly Weather Review* 96(11): 753-766.

Hurrell J.W., and J.W. Trenberth. 1999. Global sea surface temperature analyses: Multiple problems and their implications for climate analysis, modeling and reanalysis. *Bulletin of the American Meteorological Society* 80(12): 2,661-2,678.

Hurtt, G.C., L. Rosentrater, S. Frolking, and B. Moore. 2001. Linking remote-sensing estimates of land cover and census statistics on land use to produce maps of land use of the conterminous United States. *Global Biogeochemical Cycles* 15: 673-685.

Husar, R.B., J.M. Prospero, and L.L. Stowe. 1997. Characterization of tropospheric aerosols over the oceans with the NOAA advanced very high resolution radiometer optical thickness operational product. *Journal of Geophysical Research—Atmospheres* 102: 16,889-16,909.

Hutchison, C. 1998. Social science and remote sensing in famine early warning. Pp. 189-196, in *People and Pixels: Linking Remote Sensing and Social Science*, D. Liverman, E. Moran, R. Rindfuss, and P. Stern eds. National Academy Press, Washington, D.C.

Iaffaldano, G., H.-P. Bunge, and T. H. Dixon. 2006. Feedback between mountain belt growth and plate convergence. *Geology* 34(10): 893-896.

Inamdar, A.K., and V. Ramanathan. 1998. Tropical and global scale interactions among water vapor, atmospheric greenhouse effect, and surface temperature. *Journal of Geophysical Research—Atmospheres* 103: 32,177-32,194.

Isern-Fontanet, J., E. Garcia-Ladona, and J. Font. 2003. Identification of marine eddies 13 from altimetric maps. *Journal of Atmospheric and Oceanic Technologies* 20: 772-778.

Isern-Fontanet, J., E. Garcia-Ladona, and J. Font. 2006. Vortices of the Mediterranean Sea: An altimetric perspective. *Journal of Physical Oceanography* 36: 87-103.

Jacob, D.J., B.G. Heikes, S.-M. Fan, J.A. Logan, D.L. Mauzerall, J.D. Bradshaw, H.B. Singh, G.L. Gregory, R.W. Talbot, D.R. Blake, and G.W. Sachse. 1996. Origin of ozone and NO_x in the tropical troposphere: A photochemical analysis of aircraft observations over the South Atlantic basin. *Journal of Geophysical Research* 101: 24,235-24,250.

Jacquemond, S., and F. Baret, 1990. PROSPECT: A model of leaf optical properties spectra. *Remote Sensing of Environment* 44: 281-292.

Jaeglé, L., R.V. Martin, K. Chance, L. Steinberger, T.P. Kurosu, D.J. Jacob, A.I. Modi, V. Yoboué, L. Sigha-Nkamdjou, and C. Galy-Lacaux. 2004. Satellite mapping of rain-induced nitric oxide emissions from soils. *Journal of Geophysical Research* 109: D21310, doi:10.1029/2004JD004787.

Jenkins, G.S., J. Ryu, A.M. Thompson, and J.C. Witte. 2003. Linking horizontal and vertical transports of biomass fire emissions to the Tropical Atlantic Ozone Paradox during the Northern Hemisphere winter season. *Journal of Geophysical Research* 108(D23): 4745, doi:10.1029/2002JD003297.

Jenkins, G.S., and J-H. Ryu. 2004. Space-borne observations link the tropical Atlantic ozone maximum and paradox to lightning. *Atmospheric Chemistry and Physics* 4: 361-375.

Johannessen, J.A., S. Sandven, and D. Durand 2001. Oceanography. Pp. 1,585-1,622 in *The Century of Space Science*, J.A. Bleeker, J. Geiss, and M. Huber, eds. Kluwer Academic Publishers, Norwell, Ma.

Johannessen, O.M., L. Bengtsson, M.W. Miles, S.I. Kuzima, V.A. Semenov, G.V. Alekseev, A.P. Nagurnyi, V.F. Zakharov, L.P. Bobylev, L.H. Pettersson, K. Hasselmann, and H.P. Cattle. 2004. Arctic climate change—observed and modelled temperature and sea ice variability. *Tellus* (A) 56A: 1-18.

Johnston, H. 1971. Reduction in stratospheric ozone by nitrogen oxide catalysts from supersonic transport exhaust. *Science* 173: 517-522.

Jonsson, S., H. Zebker, and F. Amelung. 2005. On trapdoor faulting at Sierra Negra volcano, Galapagos. *Journal of Volcanology and Geothermal Research* 144: 59-71.

Jordan, C.F. 1969. Derivation of leaf area index from quality of light on the forest floor. *Ecology* 50: 663-666.

Joughin, I., L. Gray, R. Bindschadler, S. Price, D. Morse, C. Hulbe, K. Mattar, C. Werner. 1999. Tributaries of west Antarctic ice streams revealed by RADARSAT interferometry. *Science* 286(5438): 283-286.

Joughin, I., M. Fahnestock, D. MacAyeal, J.L. Bamber, and P. Gogineni. 2001. Observation and analysis of ice flow in the largest Greenland ice stream. *Journal of Geophysical Research—Atmospheres* 106(D24): 34,021-34,034.

Jury, M., and N. Walker. 1988. Marine boundary layer modification across the edge of the Agulhas Current. *Journal of Geophysical Research* 93: 647-654.

Justice, C.O., J.R.G. Townshend, B.N. Holben, and C.J. Tucker. 1985. Analysis of the phenology of global vegetation using meteorological satellite data. *International Journal of Remote Sensing* 6(8): 1,271-1,318.

Kahn, R.A., B.J. Gaitley, J.V. Martonchik, D.J. Diner, K.A. Crean, and B. Holben. 2005. Multiangle Imaging Spectroradiometer (MISR) global aerosol optical depth validation based on 2 years of coincident Aerosol Robotic Network (AERONET) observations. *Journal of Geophysical Research—Atmopsheres* 110: D10S04, doi:10.1029/2004JD004706.

Kalnay, E. 2003. *Atmospheric Modeling, Data Assimilation, and Predictability.* Cambridge University Press, Cambridge, UK.

Kaplan, L.D. 1959. Inference of atmospheric structure from remote radiation measurements. *Journal of the Optical Society of America.* 49: 1,004-1,007.

Kasischke, E., D. Williams, and D. Barry. 2002. Analysis of the patterns of large fires in the boreal forest region of Alaska. *International Journal of Wildland Fire* 11: 131-144.

Kaufman, Y.J., O. Boucher, D. Tanre, M. Chin, L.A. Remer, and T. Takemura. 2005. Aerosol anthropogenic component estimated from satellite data. *Geophysical Research Letters* 32: L17804, doi:10.1029/2005GL023125.

Keeling, C.D., R.B. Bacastow, A.E. Bainbridge, C.A. Ekdahl, P.R. Guenther, and L.S. Waterman. 1976. Atmospheric carbon dioxide variations at Mauna Loa Observatory, Hawaii. *Tellus* 28: 538-551.

Keeling, C.D., R.B. Bacastow, A.F. Carter, S.C. Piper, T.P. Whorf, M. Heimann, W.G. Mook, and H. Roeloffzen. 1989. A three-dimensional model of atmospheric CO_2 transport based on observed winds: 1. Analysis of observational data. In *Aspects of Climate Variability in the Pacific and the Western Americas*, D.H. Peterson, ed. *Geophysical Monograph* 55: 165-235.

Kelley, O.A., J. Stout, and J.B. Halverson. 2004. Tall precipitation cells in tropical cyclone eyewalls are associated with tropical cyclone intensification. *Geophysical Research Letters* 31: L124112, doi:10.1029/2004GL021616.

Kerr, J.T., and J. Cihlar. 2003. Land use and cover with intensity of agriculture for Canada from satellite and census data. *Global Ecology and Biogeography* 12: 161-172.

Kerr, R. 2006. A worrying trend of less ice, higher seas. *Science* 311: 1698.

Khromova, T.E., M.B. Dyurgerov, and R.G. Barry. 2003. Late-twentieth century changes in glacier extent in the Ak-shirak Range, Central Asia, determined from historical data and ASTER imagery. *Geophysical Research Letters* 30(16): 1863, doi:10.1029/2003GL017233.

Kidder, S.Q., W.M. Gray, and T.H. Vonder Haar. 1980. Tropical cyclone outer surface winds derived from satellite microwave sounder data. *Monthly Weather Review* 108: 144-152.

Kidder, S.Q., and T.H. Vonder Haar. 1995. *Satellite Meteorology: An Introduction.* Academic Press, San Diego.

Kiehl, J.T., and K.E. Trenberth. 1997. Earth's annual global mean energy budget. *Bulletin of the American Meteorological Association* 78: 197-208.

Kimball, J. S., K. C. McDonald, and M. Zhao. 2006. Spring thaw and its effect on northern terrestrial vegetation productivity observed from satellite microwave and optical remote sensing. *Earth Interactions* 10(21): 1-22.

King, M.D., W.P. Menzel, Y.J. Kaufman, D. Tanre, B.C. Gao, S. Platnick, S.A. Ackerman, L.A. Remer, R. Pincus, and P.A. Hubanks. 2003. Cloud and aerosol properties, precipitable water, and profiles of temperature and water vapor from MODIS. *IEEE Transactions on Geoscience and Remote Sensing* 41: 442-458.

Kley, D., P.J. Crutzen, H.G. J. Smit, H. Vömel, S.J. Oltmans, H. Grassl, V. Ramanathan. 1996. Observations of near-zero ozone concentrations over the convective Pacific: Effects on air chemistry. *Science* 274(5285): 230-233.

Kokaly, R.F., and R.N. Clark. 1999. Spectroscopic determination of leaf biochemistry using band-depth analysis of absorption features and stepwise multiple linear regression—an improved approach. *Remote Sensing of Environment* 67: 267-287.

König, M., J.-G. Winther, and E. Isaksson. 2001. Measuring snow and glacier ice properties from satellite. *Reviews of Geophysics* 39(1): 1-28.

Kushnir, Y., W.A. Robinson, I. Blade, N.M.J. Hall, S. Peng, and R. Sutton. 2002. Atmospheric GCM response to extratropical SST anomalies: Synthesis and evaluation. *Journal of Climate* 15: 2,233-2,256.

Lamarque, J.-F., D.P. Edwards, L.K. Emmons, J.C. Gille, O. Wilhelmi, C. Gerbig, D. Prevedel, M.N. Deeter, J. Warner, D.C. Ziskin, B. Khattatov, G.L. Francis, V. Yudin, S. Ho, D. Mao, J. Chen, and J.R. Drummond. 2003. Identification of CO plumes from MOPITT data: Application to the August 2000 Idaho-Montana forest fires. *Geophysical Research Letters* 30(13): 1688, doi:10.1029/2003GL017503.

Lambert, A., R.G. Grainger, J.J. Remedios, C.D. Rodgers, M. Corney, and F.W. Taylor. 1993. Measurements of the evolution of the Mt. Pinatubo aerosol cloud by isams. *Geophysical Research Letters* 20: 1,287-1,290.

Lamprey, H.F.. 1975. *Report on the Desert Encroachment Reconnaissance in Northern Sudan, 21 Oct. to 10 Nov., 1975.* UNESCO/UNEP.

Lamprey, H.F. 1988. Report on the desert encroachment reconnaissance in northern Sudan 21 Oct. to 10 Nov. 1975. *Desertification Control Bulletin* 17: 1-7.

Landgrebe, D. 1997. The evolution of Landsat data analysis. *Photogrammetric Engineering and Remote Sensing* 63: 859-867.

Laplace, P.S. 1776. *Recherches sur plusiers points du system du monde, Mémoures de L'Académie Royale des Sciences de Paris.* Reprinted in *Euvres complètes de Laplace*, Gauthier Villard, Paris, 9, 1893.

Laurs, R.M., P.C. Fiedler, and D.R. Montgomery. 1984. Albacore tuna catch distributions relative to environmental features observed from satellites. *Deep Sea Research* 31: 1,088-1,099.

Le Provost, C. 1983. An analysis of Seasat altimeter measurements over a coastal area: The English Channel. *Journal of Geophysical Research* 88(C3): 1,647-1,654.

Le Provost, C. 2001. Ocean tides. Pp. 267-303 in *Satellite Altimetry and Earth Sciences: A Handbook of Techniques and Applications,* L.-L. Fu and A. Cazenave, eds. Academic Press, San Diego.

Lee, T., and P. Cornillon. 1995. Temporal variation of meandering intensity and domain-wide lateral oscillations of the Gulf Stream. *Journal of Geophysical Research* 100: 13,603-13,613.

Lepers, E., E.F. Lambin, A.C. Janetos, R. DeFries, F. Achard, N. Ramankutty, and R.J. Scholes. 2005. A synthesis of information on rapid land-cover change for the period 1981-2000. *Bioscience* 55: 115-124.

Liverman, D., E. Moran, R. Rindfuss, and P. Stern (eds.) 1998. *People and Pixels: Linking Remote Sensing and Social Science*. National Academy Press, Washington, D.C.:

Livesey, N.J., L.J. Kovalenko, R.J. Salawitch, I.A. MacKenzie, M.P. Chipperfield, W.G. Read, R.F. Jarnot, and J.W. Waters. 2006. EOS Microwave Limb Sounder observations of upper stratospheric BrO: Implications for total bromine. *Geophysical Research Letters* (33): L2081, doi:10.1029/2006GL026930.

Loeb, N.G., B.A. Wielicki, F.G. Rose, and D.R. Doelling. 2007. Variability in global top-of-atmosphere shortwave radiation between 2000 and 2005. *Geophysical Research Letters* 34, L03704, doi: 10.1029/2006GL028196.

Longhurst, A., S. Sathyendranath, T. Platt, and C. Caverhill. 1995. An estimate of global primary production in the ocean from satellite radiometer data. *Journal of Plankton Research* 17: 1,245-1,271.

Lord, S.J. 2006. A historical perspective of atmospheric science from space. Presentation to the National Academies Committee on Scientific Accomplishments of Earth Observations from Space, November 10, 2006, Washington, D.C.

Loveland, T.R., B.C. Reed, J.F. Brown, D.O. Ohlen, Z. Zhu, L. Yang, and J.W. Merchant. 2000. Development of a global land cover characteristics database and IGBP DISCover from 1 km AVHRR data. *International Journal of Remote Sensing* 21: 1,303-1,330.

Luthcke, S.B., H.J. Zwally, W. Abdalati, D.D. Rowlands, R.D. Ray, R.S. Nerem, F.G. Lemoine, J.J. McCarthy, and D.S. Chinn. 2006. Recent Greenland ice mass loss by drainage system from satellite gravity observations. *Science* 314(5803): 1,286-1,289.

MacDonald, R.B., and F.G. Hall. 1980. Global crop forecasting. *Science* 208: 670-679.

Maddox, R.A. 1980. Mesoscale convective complexes. *Bulletin of the American Meteorological Society* 61: 1374-1387.

Malakoff, D. 2000. Earth-monitoring satellites: Will the U.S. bring down the curtain on Landsat? *Science* 288: 2309-2311.

Maloney, E.D., and D.B. Chelton. 2006. An assessment of the sea surface temperature influence on surface wind stress in numerical weather prediction and climate models. *Journal of Climate* 19: 2,743-2,762.

Martin J.H., K.H. Coale, K.S. Johnson, S.E. Fitzwater, R.M. Gordon, S.J. Tanner, C.N. Hunter, V.A. Elrod, J.L. Nowicki, T.L. Coley, R.T. Barber, S. Lindley, A.J. Watson, K. Van Scoy, C.S. Law, M.I. Liddicoat, R. Ling, T. Stanton, J. Stockel, C. Collins, A. Anderson, R. Bidigare, M. Ondrusek, M. Latasa, F.J. Millero, K. Lee, W. Yao, J.Z. Zhang, G. Friederich, C. Sakamoto, F. Chavez, K. Buck, Z. Kolber, R. Greene, P. Falkowski, S.W. Chisholm, F. Hoge, R. Swift, J. Yungel, S. Turner, P. Nightingale, A. Hatton, P. Liss, and N.W. Tindal. 1994. Testing the iron hypothesis in ecosystems of the equatorial Pacific Ocean. *Nature* 371: 123-129.

Massonnet, D., M. Rossi, C. Carmona, F. Adragna, G. Peltzer, K. Feigl, and T. Rabaute. 1993. The displacement field of the Landers earthquake mapped by radar interferometry. *Nature* 364: 138-142.

Matsuno, T. 1971. *A Dynamical Model of the Stratospheric Sudden Warming*. Geophysical Fluid Dynamics Laboratory, NOAA, Princeton University, Princeton, N.J.

Matthews, E. 1983. Global vegetation and land use: New high resolution data bases for climate studies. *Journal of Climatology and Applied Meteorology* 22: 474-487.

Maxwell, E.L. 1976. Multivariate system analysis of multispectral imagery. *Photogrammetric Engineering and Remote Sensing* 42: 1173-1186.

McCarthy, J.J. 1999. The evolution of the Joint Global Ocean Flux Study project. In *The Changing Ocean Carbon Cycle: A Midterm Synthesis of the Joint Global Ocean Flux Study*, R.B. Hanson, H.W. Ducklow, and J.G. Field, eds. Cambridge University Press, New York.

McCauley, J.F., G.G. Schaber, C.S. Breed, M.J. Grolier, C.V. Haynes, B. Issawi, C. Elachi, and R. Blom. 1982. Subsurface valleys and geoarchaeology of the eastern Sahara revealed by Shuttle Radar. *Science* 218: 1004-1020.

McCormick, M.P., H.M. Steele, P. Hamill, W.P. Chu, and T.J. Swissler. 1982. Polar stratospheric cloud sightings by SAMII. *Journal of Atmospheric Science* 39: 1,387-1,397.

McCormick, M.P. 1992. SAGE II measurements of early Pinatubo aerosols. *Geophysical Research Letters* 19: 155-158.

McCormick, M.P., P.-H. Wang, and L.R. Poole. 1993. Stratospheric aerosols and clouds. Pp. 205-219 in *Aerosol-Cloud-Climate Interactions*, P.V. Hobbs, ed. Academic Press, New York.

McCormick, M.P., L.W. Thomason, and C.R. Trepte. 1995. Atmospheric effects of the Mt. Pinatubo eruption. *Nature* 373: 399-404.

McElroy, M.B., R.J. Salawitch, S.C. Wofsy, and J.A. Logan. 1986. Reductions of Antarctic ozone due to synergistic interactions of chlorine and bromine. *Nature* 321: 759-762.

McGuire, M., A.W. Wood, A.F. Hamlet, and D.P. Lettenmaier. 2006. Use of satellite data for streamflow and reservoir storage forecasts in the Snake River basin. *Journal of Water Resources Planning and Management* 132(2): 97-110.

McIver, D.K., and M.A. Friedl. 2002. Using prior probabilities in decision-tree classification of remotely sensed data. *Remote Sensing of the Environment* 81: 253-261.

McMillan, W.W., C. Barnet, L. Strow, M.T. Chahine, M.L. McCourt, P.C. Novelli, S. Korontzi, E.S. Maddy, and S. Datta. 2005. Daily global maps of carbon monoxide: First views from NASA's Atmospheric InfraRed Sounder. *Geophysical Research Letters* 32: L11801, doi:10.1029/2004GL021821.

Melbourne, T.I., and F.H Webb. 2003. Slow but not quite silent. *Science* 300(5,627): 1,886-1,887.

Mergenthaler, J., J.B. Kumer, A.E. Roche, R.W. Nightingale, J.F. Potter, P.L. Bailey, S.T. Massie, J.C. Gille, D.P. Edwards, M.R. Gunson, M.C. Abrams, G.C. Toon, B. Sen, J.F. Blavier, P.S. Connell, D.E. Kinnison, D.G. Murcray, F.H. Murcray, A. Goldman, and E.C. Zipf. 1996. Validation of CLAES $ClONO_2$ measurements. *Journal of Geophysical Research* 101: 9,603-9,620.

Milliff, R.F., J. Morzel, D.B. Chelton, and M.H. Freilich. 2004. Wind stress curl and wind stress divergence biases from rain effects on QSCAT surface wind retrievals. *Journal of Atmospheric and Oceanic Technology* 21: 1,216-1,231.

Minnis, P., E.F. Harrison, L.L. Stowe, G.G. Gibson, F.M. Denn, D.R. Doelling, and W.L. Smith. 1993. Radiative climate forcing by the Mount Pinatubo eruption. *Science* 259: 1,411-1,415.

Mishchenko, M.I., I.V. Geogdzhayev, B. Cairns, W.B. Rossow, and A.A. Lacis. 1999. Aerosol retrievals over the ocean by use of channels 1 and 2 AVHRR data: Sensitivity analysis and preliminary results. *Applied Optics* 38: 7,325-7,341.

Molina, M.J., and F.S. Rowland. 1974. Stratospheric sink for chlorofluoromethanes: Chlorine atomc-atalysed destruction of ozone. *Nature* 249: 810-812.

Molotch, N.P., T.H. Painter, R.C. Bales, and J. Dozier. 2004. Incorporating remotely-sensed snow albedo into a spatially-distributed snowmelt model. *Geophysical Research Letters* 31(3): L03501, doi:10.1029/2003GL019063.

Mooney, H.A. 1991. Emergence of the study of global ecology: Is terrestrial ecology an impediment to progress? *Ecological Applications* 1: 2-5.

Morel, A., and L. Prieur. 1977. Analysis of variations in ocean color. *Limnology and Oceanography* 22(4): 709-722.

Morris, G.A., S. Hersey, A.M. Thompson, A. Stohl, P.R. Colarco, W.W. McMillan, J. Warner, B.J. Johnson, J.C. Witte, T.L. Kucsera, D.E. Larko, and S.J. Oltmans. 2006. Alaskan and Canadian forest fires exacerbate ozone pollution in Houston, Texas, on 19 and 20 July 2004. *Journal of Geophysical Research* 111: D24S03, doi: 2006JD007090.

Morrow, R., F. Birol, D. Griffin, and J. Sudre. 2004. Divergent pathways of cyclonic and anti-cyclonic ocean eddies. *Geophysical Research Letters* 31: L24311, doi:10.1029/2004GL020974.

Mote, P.W., K.H. Rosenlof, M.E. McIntyre, E.S. Carr, J.C. Gille, J.R. Holton, J.S. Kinnersley, H.C. Pumphrey, J.M. Russell, and J.W. Waters. 1996. An atmospheric tape recorder: The imprint of tropical tropopause temperatures on stratospheric water vapor. *Journal of Geophysical Research* 101: 2,989-4,006.

Moxim, W.J., and H. Levy. 2000. A model analysis of the tropical South Atlantic Ocean tropospheric ozone maximum: The interaction of transport and chemistry. *Journal of Geophysical Research* 105(D13): 17,393-17,416.

Muller-Karger, F.E., C.R. McClain, and P.R. Richardson. 1988. Dispersal of the Amazon's water. *Nature* 333(6168): 56-59.

Munk, W., and C. Wunsch. 1998. Abyssal recipes II: Energetics of tidal and wind mixing. *Deep-Sea Research* 45: 1,976-2,009.

Myneni, R.B., C.D. Keeling, C.J. Tucker, G. Asrar, and R.R. Nemani. 1997. Increased plant growth in the northern high latitudes from 1981 to 1991. *Nature* 386(6626): 698-702.

Myneni, R.B., C.J. Tucker, G. Asrar, and C.D. Keeling. 1998. Interannual variations in satellite-sensed vegetation index data from 1981 to 1991. *Journal of Geophysical Research—Atmospheres* 103(D6): 6,145-6,160.

Nakajima, T., and A. Higurashi. 1998. A use of two-channel radiances for an aerosol characterization from space. *Geophysical Research Letters* 25: 3,815-3,818.

NASA (National Aeronautics and Space Administration). 1987. *Leadership and America's Future in Space.* NASA, Washington, D.C.

NASA. 2005. *Tropical Rainfall Measuring Mission (TRMM) Senior Review Proposal 2005.* Goddard Space Flight Center, Greenbelt, MD.

Nightingale, R.W., A.E. Roche, J.B. Kumer, J.L. Mergenthaler, J.C. Gille, S.T. Massie, P.L. Bailey, D.P. Edwards, M.R. Gunson, G.C. Toon, B. Sen, and P.S. Connell. 1996. Global CF_2C_{12} measurements by UARS cryogenic limb array etalon spectrometer: Validation by correlative data and a model. *Journal of Geophysical Research* 101(D6): 9,711-9,736

NRC (National Research Council). 1979. *Protection against Depletion of Stratospheric Ozone by Chlorofluorocarbons.* National Academy Press, Washington, D.C.

NRC. 1985. *A Strategy for Earth Science from Space in the 1980s and 1990s: Part II: Atmosphere and Interactions with the Solid Earth, Oceans, and Biota.* National Academy Press, Washington, D.C.

NRC 1999. *Adequacy of Climate Observing Systems.* National Academy Press. Washington, D.C.

NRC. 2000. *Issues in the Integration of Research and Operational Satellite Systems for Climate Research. Part 1: Science and Design.* National Academy Press, Washington, D.C.

NRC. 2001a. *Under the Weather: Climate, Ecosystems, and Infectious Disease.* National Academy Press, Washington, D.C.

NRC. 2001b. *Improving the Effectiveness of U.S. Climate Modeling.* National Academy Press, Washington, D.C.

NRC. 2002. *Toward New Partnerships in Remote Sensing: Government, the Private Sector, and Earth Science Research.* National Academy Press, Washington, D.C.

NRC. 2003. *Satellite Observations of the Earth's Environment: Accelerating the Transition of Research to Operations.* The National Academies Press, Washington, D.C.

NRC. 2004. *Climate Data Records from Environmental Satellites.* The National Academies Press, Washington, D.C.

NRC. 2005. *Earth Science and Application from Space: Urgent Needs and Opportunities to Serve the Nation.* The National Academies Press, Washington, D.C..

NRC. 2006. *Assessment of the Benefits of Extending the Tropical Rainfall Measuring Mission: A Perspective from the Research and Operations Communities, Interim Report.* The National Academies Press, Washington, D.C.

NRC. 2007a. *Earth Science and Applications from Space: National Imperatives for the Next Decade and Beyond.* The National Academies Press, Washington, D.C..

NRC. 2007b. *Strategic Guidance for the National Science Foundation's Support of the Atmospheric Sciences.* The National Academies Press, Washington, D.C.

NRC. 2007c. *Polar Icebreakers in a Changing World: An Assessment of U.S. Needs.* The National Academies Press, Washington, D.C..

NWS (National Weather Service). 2006. History of the National Weather Service. Available at http://www.weather.gov/pa/history/index.php. Accessed February 23, 2007.

O'Keefe, J.A., A. Eckels, and R.K. Squires. 1960. Pear-shaped component of the geoid from the motion of vanguard 1. *Annals of the International Geophysical Year* 12(1): 199-201.

Oldeman, L.R., R.T.A. Hakkeling, and W.G. Sombroek. 1991. *World Map of the Status of Human Induced Soil Degradation: An Explanatory Note.* International Soil Reference and Information Centre, Wageningen, the Netherlands.

Olson, D.B. 2001. Biophysical dynamics of western transition zones: A preliminary synthesis. *Fisheries Oceanography* 10: 133-150.

Olson, J.S., J.A. Watts, and L.J. Allison. 1983. *Carbon in Live Vegetation of Major World Ecosystems.* Oak Ridge National Laboratory, Oak Ridge, Tenn.

O'Neill, L.W., D.B. Chelton, and S.K. Esbensen. 2003. Observations of SST-induced perturbations of the wind stress field over the Southern Ocean on seasonal time scales. *Journal of Climate* 16: 2,340-2,354.

O'Reilly, J.E., S. Maritorena, M.C. O'Brien, D.A. Siegel, D. Toole, D. Menzies, R.C. Smith, J.L. Mueller, B.G. Mitchell, M. Kahru, F.P. Chavez, P. Strutton, G.F. Cota, S.B. Hooker, C. McClain, K.L. Carder, F. Muller-Karger, L. Harding, A. Magnuson, D. Phinney, G.F. Moore, J. Aiken, K.R. Arrigo, R. Letelier, and M. Culver. 2000. Ocean color chlorophyll *a* algorithms for SeaWiFS, OC2, and OC4: Version 4. Pp. 9-23, in *SeaWiFS Postlaunch Calibration and Validation Analyses, Part 3*, S.B. Hooker and E.R. Firestone, eds. NASA Tech. Memo. 2000-206892, Vol. 11. NASA Goddard Space Flight Center, Greenbelt, Md.

Painter, T.H., J. Dozier, D.A. Roberts, R.E. Davis, and R.O. Green. 2003. Retrieval of subpixel snow-covered area and grain size from imaging spectrometer data. *Remote Sensing of Environment* 85(1): 64-77.

Palacios-Orueta, A., A. Parra, E. Chuvieco, and C. Carmona-Moreno. 2004. Remote sensing and geographic information systems methods for global spatiotemporal modeling of biomass burning emissions: Assessment in the African continent. *Journal of Geophysical Research* 109: D14S09, doi:10.1029/2004JD004734.

Parkinson, C.L., D.J. Cavalieri, P. Gloersen, H.J. Zwally, and J.C. Comiso. 1999. Arctic sea ice extents, areas, and trends, 1978-1996. *Journal of Geophysical Research* 104(C9): 20,837-20,856.

Pearson, R.L., and L.D. Miller. 1972. Remote mapping of standing crop biomass for estimation of the productivity of the shortgrass prairie. Pp. 1357-1381 in *8th International Symposium on Remote Sensing of Environment.* University of Michigan, Ann Arbor.

Pereira J., B. Pereira, P. Barbosa, D. Stroppiana, M. Vasconcelos, and J. Gregoire. 1999. Satellite monitoring of fire in the EXPRESSO study area during the 1996 dry season experiment: Active fires, burnt area, and atmospheric emissions. *Journal of Geophysical Research—Atmospheres* 104: 30,701-30,712.

Pierce, R.B., T.K. Schaack, J. Al-Saadi, T.D. Fairlie, C. Kittaka, G. Lingenfelser, M. Natarajan, J. Olson, A. Soja, T.H. Zapotocny, A. Lenzen, J. Stobie, D.R. Johnson, M. Avery, G. Sachse, A. Thompson, R. Cohen, J. Dibb, J. Crawford, D. Rault, R. Martin, J. Szykman, and J. Fishman. 2007. Chemical data assimilation estimates of continental U.S. ozone and nitrogen budgets during INTEX-A. *Journal of Geophysical Research—Atmospheres* 112: D12S18, doi:10.1029/2006JD007722.

Pimentel, D., R. Zuniga, and D. Morrison. 2005. Update on the environmental and economic costs associated with alien-invasive species in the United States. *Ecological Economics* 52: 273-288.

Platt, T., and S. Sathyendranath. 1988. Ocean primary production-estimation by remote sensing at local and regional scales. *Science* 241: 1,613-1,620.

Platzman, G.W. 1981. Normal modes of the world ocean. II. Description of modes in the period range 8 to 80 hours. *Journal of Physical Oceanography* 11: 579-603.

Potter, C.S., J.T. Randerson, C.B. Field, P.A. Matson, P.M. Vitousek, H.A. Mooney, and S.A. Klooster. 1993. Terrestrial ecosystem production: A process model based on global satellite and surface data. *Global Biogeochemical Cycles* 12: 1,313-1,330.

Prince, S.D., E.B. De Colstoun, and L.L. Kravitz. 1998. Evidence from rain-use efficiencies does not indicate extensive Sahelian desertification. *Global Change Biology* 4: 359-374.

Prospero, J.M., P. Ginoux, O. Torres, S.E. Nicholson, and T.E. Gill. 2002. Environmental characterization of global sources of atmospheric soil dust identified with the Nimbus 7 Total Ozone Mapping Spectrometer (TOMS) absorbing aerosol product. *Review of Geophysics* 40(1): 1002, doi:10.1029/2000RG000095.

PSAC (President's Science Advisory Committee). 1958. *Introduction to Outer Space*. U.S. Government Printing Office, Washington, D.C.

Purdom, J.F.W. 1976. Some uses of high-resolution GOES imagery in the mesoscale forecasting of convection and its behavior. *Monthly Weather Review* 104: 1,474-1,483.

Purdom, J.F.W. 1986. Convective Scale Interaction: Arc Cloud Lines and the Development and Evolution of Deep Convection. Ph.D. dissertation, Department of Atmospheric Science, Colorado State University, Fort Collins.

Ramanathan, V., R.D. Cess, E.F. Harrison, P. Minnis, B.R. Barkstrom, E. Ahmad, and D. Hartmann. 1989. Cloud-radiative forcing and climate: Results from the Earth Radiation Budget Experiment. *Science* 243: 57-63.

Ramankutty, N., and J.A. Foley. 1999. Estimating historical changes in global land cover: Croplands from 1700 to 1992. *Global Biogeochemical Cycles* 13(4): 997-1027.

Ramankutty, N., H.K. Gibbs, F. Achard, R. DeFries, J.A. Foley, and R.A. Houghton. 2007. Challenges to estimating carbon emissions from tropical deforestation. *Global Change Biology* 13: 51-66.

Ramankutty, N., A. Evan, C. Monfreda, and J.A. Foley. In press. Farming the Planet. 1: The Geographic Distribution of Global Agricultural Lands in the Year 2000. *Global Biogeochemical Cycles*.

Rankin, A.M., E.W. Wolff, and S. Martin. 2002. Frost flowers: Implications for tropospheric chemistry and ice core interpretation. *Journal of Geophysical Research* 107(D23): 4683, doi:10.1029/2002JD002492.

Rasmusson, E.M., and T.H. Carpenter. 1982. Variations in tropical sea surface temperature and surface wind fields associated with the Southern Oscillation/El Niño. *Monthly Weather Review* 110: 354-384.

Raval, A., and V. Ramanathan. 1989. Observational determination of the greenhouse effect. *Nature* 342: 758-761.

Ray, R.D., and G.T. Mitchum. 1997. Surface manifestation of internal tides in the deep ocean. *Progress in Oceanography* 40: 135-162.

Redman, C. 1999. *Human Impact on Ancient Environments*. University of Arizona Press, Tucson.

Reigber, C.H., P. Schwintzer, K.-H. Neumayer, F. Barthelmes, R. König, Ch. Förste, G. Balmino, R. Biancale, J.-M. Lemoine, S. Loyer, S. Bruinsma, F. Perosanz, and T. Fayard. 2003. The CHAMP-only Earth Gravity Field Model EIGEN-2. *Advances in Space Research* 31: 1,883-1,888.

Remer, L.A., D. Tanre, Y.J. Kaufman, C. Ichoku, S. Mattoo, R. Levy, D.A. Chu, B. Holben, O. Dubovik, A. Smirnov, J.V. Martins, R.R. Li, and Z. Ahmad. 2002. Validation of MODIS aerosol retrieval over ocean. *Geophysical Research Letters* 29(12): 8008, doi 10.1029/2001GL013204.

Remer, L.A., Y.J. Kaufman, D. Tanre, S. Mattoo, D.A. Chu, J.V. Martins, R.R. Li, C. Ichoku, R.C. Levy, R.G. Kleidman, T.F. Eck, E. Vermote, and B.N. Holben. 2005. The MODIS aerosol algorithm, products, and validation. *Journal of Atmospheric Science* 62: 947-973.

Reynolds, J.F., and D.M. Stafford-Smith, eds. 2002. *Global Desertification: Do Humans Cause Deserts?* Dahlem University Press, Berlin.

Reynolds, J.F., D.M.S. Smith, E.F. Lambin, B.L. Turner II, M. Mortimore, S.P.J. Batterbury, T.E. Downing, H. Dowlatabadi, R.J. Fernandez, J.E. Herrick, E. Huber-Sannwald, H. Jiang, R. Leemans, T. Lynam, F.T. Maestre, M. Ayarza, and B. Walker. 2007. Global desertification: Building a science for dryland development. *Science* 316: 847-851.

Riaño, D., J.A. Moreno Ruiz, D. Isidoro, and S.L. Ustin. 2007. Global spatial patterns and temporal trends of burned area between 1981 and 2000 using NOAA-NASA Pathfinder. *Global Change Biology* 13: 40-50.

Richter, A., F. Wittrock, M. Eisinger, and J.P. Burrows. 1998. GOME observations of tropospheric BrO in northern hemisphere spring and summer 1997. *Geophysical Research Letters* 25: 2,683-2,686.

Richter, A., and J.P. Burrows. 2002. Tropospheric NO_2 from GOME measurements. *Advances in Space Research* 29: 1,673-1,683.

Rignot, E. 2001. Evidence for rapid retreat and mass loss of Thwaites Glacier, West Antarctica. *Journal of Glaciology* 47(157): 213-222.

Rignot, E., and P. Kanagaratnam. 2006. Changes in the velocity structure of the Greenland ice sheet. *Science* 311(5763): 986-990.

Rind, D., E.W. Chiou, W. Chu, J. Larsen, S. Oltmans, J. Lerner, M.P. McCormick, and L. McMaster. 1991. Positive water-vapor feedback in climate models confirmed by satellite data. *Nature* 349: 500-503.

Rinsland, C.P., M. Luo, J.A. Logan, R. Beer, H.M. Worden, J.R. Worden, K. Bowman, S.S. Kulawik, D. Rider, G. Osterman, M. Gunson, A. Goldman, M. Shephard, S.A. Clough, C. Rodgers, M. Lampel, and L. Chiou. 2006. Nadir measurements of carbon monoxide distributions by the Tropospheric Emission Spectrometer onboard the Aura spacecraft: Overview of analysis approach and examples of initial results. *Geophysical Research Letters* 33: L22806, doi 10.1029/2006GL027000.

Ritchie, J.C., C.H. Eyles, and C.V. Haynes. 1985. Sediment and pollen evidence for an early to mid-Holocene humid period in the eastern Sahara. *Nature* 314: 352-355.

Robinson, I.S. 1985. *Satellite Oceanography: An Introduction for Oceanographers and Remote-sensing Scientists*. John Wiley & Sons, Hoboken, NJ.

Rodell, M., P.R. Houser, U. Jambor, J. Gottschalck, K. Mitchell, C.J. Meng, K. Arsenault, B. Cosgrove, J. Radakovich, M. Bosilovich, J.K. Entin, J.P. Walker, D. Lohmann, and D. Toll. 2004. The global land data assimilation system. *Bulletin of the American Meteorological Society* 85(3): 381-394.

Rossby, C.-G et al. 1939. Relation between variations in the intensity of the zonal circulation of the atmosphere and the displacements of the semi-permanent centers of action. *Journal of Marine Research* 2: 38-55.

Rossby, C.-G. 1940. Planetary flow patterns in the atmosphere. *Quarterly Journal of the Royal Meteorological Society* 66: 68-87.

Rossow, W.B., and R.A. Schiffer. 1999. Advances in understanding clouds from ISCCP. *Bulletin of the American Meteorology Society* 80: 2,261-2,287.

Roughgarden, J., Running, S.W., Matson, P.A. 1991. What does remote-sensing do for ecology? *Ecology* 72(6): 1,918-1,922.

Rouse, J.W., R.H. Haas, J.A. Schell, and D.W. Deering. 1973. Monitoring vegetation systems in the great plains with ERTS. 3rd ERTS Symposium, NASA SP-351 I: 309-317.

Rouse, J.W., R.H. Haas, J.A. Schell, D.W. Deering, and J.C. Harlan. 1974. Monitoring the vernal advancement and retrogradation (greenwave effect) of natural vegetation. NASA/GSFC Type III Final Report. NASA, Greenbelt, Md.

Rowntree, P.R. 1972. The influence of tropical east Pacific Ocean temperatures on the atmosphere. *Quarterly Journal of the Royal Meteorological Society* 98: 290-321.

Running, S.W., and E.R. Hunt. 1993. Generalization of a forest ecosystem process model for other biomes, BIOME-BGC, and an application for global scale models. Pp. 141-158 in *Scaling Physiological Processes: Leaf to Globe*, J.R. Ehleringer and C.B. Field, eds. Academic Press, San Diego.

Russell, J.M., M.Z. Luo, R.J. Cicerone, and L.E. Deaver. 1996. Satellite confirmation of the dominance of chloroflourocarbons in the global stratospheric chlorine budget. *Nature* 379: 526-529.

Salisbury, J.E., J.W. Campbell, L.D. Meeker, and C.J. Vörösmarty. 2001. Ocean color and river data reveal fluvial influence in coastal waters. *AGU-EOS Transactions* 82(20): 221, 226-227.

Salisbury, J.S, J.W. Campbell, E.L. Linder, L.D. Meeker, F.E. Muller-Karger, and C.J. Vorosmarty. 2004. On the seasonal correlation of surface particle fields with wind stress and Mississippi discharge in the northern Gulf of Mexico. *Deep Sea Research Part II* 51(10-11): 1,187-1,203.

Sauvage, B., V. Thouret, A.M. Thompson, J.C. Witte, J.-P. Cammas, P. Nedelec, and G. Athier. MOZAIC and SHADOZ Teams. 2006. Enhanced view of the "Tropical Atlantic Ozone Paradox" and "Zonal Wave-one" from the in-situ MOZAIC and SHADOZ data. *Journal of Geophysical Research* 111: D01301, doi: 10.129/2005JD006241.

Schiffer, R.A., and W.B. Rossow. 1985. ISCCP global radiance data set: A new resource for climate research. *Bulletin of the American Meteorological Society* 66: 1,498-1,505.

Schneider, A., M.A. Friedl, D.K. McIver, and C.E. Woodcock. 2003. Mapping urban areas by fusing multiple sources of coarse resolution remotely sensed data. *Photogrammetric Engineering and Remote Sensing* 69: 1,377-1,386.

Schneider, A., and C.E. Woodcock. In press. Compact, dispersed, fragmented, extensive? A comparison of urban growth in 25 global cities using remotely sensed data, pattern metrics and census information. *Urban Studies*.

Schollaert, S.E., T. Rossby, and J.A. Yoder. 2004. Gulf Stream cross-frontal exchange: Possible mechanisms to explain inter-annual variations in phytoplankton chlorophyll in the Slope Sea during the SeaWiFS years. *Deep-Sea Research Part II* 51: 173-188.

Schultz, M.G. 2002. On the use of ATSR fire count data to estimate the seasonal and interannual variability of vegetation fire emissions. *Atmospheric Chemistry Physics Discussions* 2: 1,159-1,179.

Sellers, P.J., C.J. Tucker, G.J. Collatz, S.O. Los, C.O. Justice, D.A. Dazlich, and D.A. Randall. 1994. A global 1-degree-by-1-degree NDVI data set for climate studies, Part 2: The generation of global fields of terrestrial biophysical parameters from the NDVI. *International Journal of Remote Sensing* 15(17): 3,519-3,545.

Sheets, R.C. 1990. The National Hurricane Center—Past, present, and future. *Weather Forecasting* 5: 185-232.

Shukla, J., and J.M. Wallace. 1983. Numerical simulation of the atmospheric response to equatorial Pacific sea surface temperature anomalies. *Journal of Atmospheric Science* 40: 1,613-1,630.

Siegel, D.A. 2001. Oceanography—The Rossby rototiller. *Nature* 409: 576-577.

Simmons, A.J., and A. Hollingsworth. 2002. Some aspects of the improvement in skill of numerical weather prediction. *Quarterly Journal of the Royal Meteorological Society* 128: 647-678.

Simon, M., S. Plummer, F. Fierens, J.J. Hoeltzemann, and O. Arino. 2004. Burnt area detection at global scale using ATSR-2: The GlobScar products and their qualification. *Journal of Geophysical Research* 109: D14S02, doi:10.1029/2003JD003622.

Simons, M., and B.H. Hager. 1997. Localization of gravity and the signature of glacial rebound. *Nature* 390: 500-504.

Simpson, J., R.F. Adler, and G. North. 1988. A proposed Tropical Rainfall Measuring Mission (TRMM) satellite. *Bulletin of the American Meteorological Society* 69: 278-295.

Simpson, J., C. Kummerow, W.-K. Tao, and R.F. Adler. 1996. On the Tropical Rainfall Measuring Mission (TRMM) satellite. *Meteorology and Atmospheric Physics* 60: 19-36.

Sinnhuber, B.-M., A. Rozanov, N. Sheode, O.T. Afe, A. Richter, M. Sinnhuber, F. Wittrock, and J.P. Burrows. 2005. Global observations of stratospheric bromine monoxide from SCIAMACHY. *Geophysical Research Letters* 32: L20810, doi: 10.1029/2005GL023839.

Skole, D., and C. Tucker. 1993. Tropical deforestation and habitat fragmentation in the Amazon: Satellite data from 1978 to 1988. *Science* 260: 1,905-1,910.

Smith, A.K., and P.L. Bailey. 1985. Comparison of horizontal winds from the LIMS satellite instrument with rocket measurements. *Journal of Geophysical Research* 20: 3,897-3,901.

Smith, B., and D. Sandwell. 2003. Accuracy and resolution of shuttle radar topography mission data. *Geophysical Research Letters* 30(9): doi:10.1029/2002GL016643.

Smith, S.E. 1986. Drought and water management: The Egyptian response. *Journal of Soil and Water Conservation* 41: 297-300.

Smith, W.L. 1970. Iterative solution of the radiative transfer equation for temperature and absorbing gas profile of an atmosphere. *Applied Optics* 9: 1,993-1,999.

Soden, B.J., R.T. Wetherald, G.L. Stenchikov, and A. Robock. 2002. Global cooling after the eruption of Mount Pinatubo: A test of climate feedback by water vapor. *Science* 296: 727-730.

Solomon S. 1999. Stratospheric ozone depletion: a review of concepts and history. *Reviews of Geophysics* 37: 275-316.

Song, T.-R. A., and M. Simons. 2003. Large trench-parallel gravity variations predict seismogenic behavior in subduction zones. *Science* 301(5633): 630-633.

Stajner, I., K. Wargan, L.-P. Chang, H. Hayashi, S. Pawson, and H. Nakajima. 2006. Assimilation of ozone profiles from the Improved Limb Atmospheric Spectrometer—II: Study of Antarctic ozone. *Journal of Geophysical Research* 111: D11S14, doi:10.1029/2005JD006448.

Steeman-Nielsen, E., and E.A. Jensen. 1957. The autotrophic production of organic matter in the oceans. *Galathea Report* I: 49-124.

Stephens, G.L., D.G. Vane, R.J. Boain, G.G. Mace, K. Sassen, Z.E. Wang, A.J. Illingworth, E.J. O'Connor, W.B. Rossow, S.L. Durden, S.D. Miller, R.T. Austin, A. Benedetti, and C. Mitrescu. 2002. The cloudsat mission and the a-train—A new dimension of space-based observations of clouds and precipitation. *Bulletin of the American Meteorological Society* 83: 1,771-1,790.

Stommel, H. 1965. *The Gulf Stream: A Physical and Dynamical Description*, 2nd ed. University of California Press, Berkeley, and Cambridge University Press, London.

Stowe, L.L., R.M. Carey, and P.P. Pellegrino. 1992. Monitoring the Mt. Pinatubo aerosol layer with NOAA-11 AVHRR data. *Geophysical Research Letters* 19: 159-162.

Stowe, L.L., A.M. Ignatov, and R.R. Singh. 1997. Development, validation, and potential enhancements to the second-generation operational aerosol product at the national environmental satellite, data, and information service of the national oceanic and atmospheric administration. *Journal of Geophysical Research—Atmospheres* 102: 16,923-16,934.

Stroeve, J.C., M.C. Serreze, F. Fetterer, T. Arbetter, W. Meier, J. Maslanik, and K. Knowles. 2005. Tracking the Arctic's shrinking ice cover: Another extreme September minimum in 2004. *Geophysical Research Letters* 32(4): L04501, doi: 10.1029/2004GL021810.

Strub, P.T., P.M. Kosro, and A. Huyer. 1991. The nature of the cold filaments in the California current system. *Journal of Geophysical Research (C) Oceans* 96(8): 14,743-14,768.

Suliman, M.M. 1988. Dynamics of range plants and desertification monitoring in the Sudan. *Desertification Control Bulletin* 16: 27-31.

Suomi, V.E. 1961. Earth's thermal radiation balance; preliminary results from Explorer 7. EOS. *Transactions of the American Geophysical Union* 42(4): 467-474.

Suomi, V.E., T.H. Vonder Haar, R.J. Krauss, and A.J. Stamm. 1971. Possibilities for sounding the atmosphere from a geosynchronous spacecraft. *Space Research* 11: 609-617.

Sweet, W.R., R. Fett, J. Kerline, and P. LaViolette. 1981. Air-sea interaction effects in the lower troposphere across the north wall of the Gulf Stream. *Monthly Weather Review* 109: 1,042-1,052.

Szeliga, W., T.I. Melbourne, M.M. Miller, and V.M. Santillan. 2004. Southern Cascadia episodic slow earthquakes. *Geophysical Research Letters* 31: L16602, doi:10.1029/2004GL020824.

Takashima, H., and M. Shiotani. 2007. Ozone variation in the tropical tropopause layer as seen from ozonesonde data. *Journal of Geophysical Research* 112: D11123, doi:10.1029/2006JD008322.

Tamisiea, M.E., J.X. Mitrovica, J.L. Davis. 2007. GRACE Gravity Data Constrain Ancient Ice Geometries and Continental Dynamics over Laurentia. *Science* 316: 881.

Tanré, D., F.M. Breon, J.L. Deuze, M. Herman, P. Goloub, F. Nadal, and A. Marchand. 2001. Global observation of anthropogenic aerosols from satellite. *Geophysical Research Letters* 28: 4,555-4,558.

Tansey, K., J.-M. Grégoire, D. Stroppiana, A. Sousa, J. M.N. Silva, J.M.C. Pereira, L. Boschetti, M. Maggi, P.A. Brivio, R. Fraser, S. Flasse, D. Ershov, E. Binaghi, D. Graetz, and P. Peduzzi. 2004. Vegetation burning in the year 2000: Global burned area estimates from SPOT VEGETATION data. *Journal of Geophysical Research—Atmospheres* 109: D14S03, doi:10.1029/2003JD003598.

Tarboton, D.G. 1997. A new method for the determination of flow directions and upslope areas in grid digital elevation models. *Water Resources Research* 33(2): 309-320.

Tardin, A.T., D.C.L. Lee, R.J.R. Santos, O.R. Assis, M.P.S. Barbosa, M.L. Moreira, M.T. Pereira, D. Silva, and C.P. Santos Filho. 1980. *Subprojeto Desmatamento, Convêino IBDF/CNPq-INPE 1979*. Technical Report INPE-1649-RPE/103. Instituto Nacional de Pesquisas Espaciais (INPE), São José dos Campos, São Paulo.

Tardin, A.T., and R.P. Cunha. 1989. *Avaliação da alteração da cobertura florestal na Amazônia Legal utilizando Sensoriamento Remoto Orbital.* Technical Report INPE-5010-RPE/607, Instituto Nacional de Pesquisas Espaciais, São José dos Campos.

Thompson, A.M., B.G. Doddridge, J.C. Witte, R.D. Hudson, W.T. Luke, J.E. Johnson, B.J. Johnson, S.J. Oltmans, and R. Weller. 2000. A tropical Atlantic paradox: Shipboard and satellite views of a tropospheric ozone maximum and wave-one in January-February 1999. *Geophysical Research Letters* 27: 3,317-3,320.

Thompson, A.M., J.C. Witte, R.D. Hudson, H. Guo, J.R. Herman, and M. Fujiwara. 2001. Tropical tropospheric ozone and biomass burning. *Science* 291: 2,128-2,132.

Thompson, A.M., J.C. Witte, R.D. McPeters, S.J. Oltmans, F.J. Schmidlin, J.A. Logan, M. Ujiwara, V.W. J.H. Kirchhoff, and G. Labow. 2003. Southern Hemisphere Additional Ozonesondes (SHADOZ) 1998-2000 tropical ozone climatology 2. Tropospheric variability and the zonal wave-one. *Journal of Geophysical Research* 108(D2): 8,238, doi:10.1029/2002JD002241.

Tierney C.C., M.E. Parke, and G.H. Born. 1998. Ocean tides from along track altimetry. *Journal of Geophysical Research* 103: 10,273-10,287.

Tilford, S. 1984. Global habitability and Earth remote sensing. *Philosophical Transactions of the Royal Society of London. Series A, Mathematical and Physical Sciences* 312(1519): 115-118.

Torres, O., P.K. Bhartia, J.R. Herman, A. Sinyuk, P. Ginoux, and B. Holben. 2002. A long-term record of aerosol optical depth from TOMS observations and comparison to AERONET measurements. *Journal of Atmospheric Science* 59: 398-413.

Townshend, J.R.G., T.E. Goff, and C.J.Tucker. 1985. Multitemporal dimensionality of images of Normalized Difference Vegetation Index at continental scales. *IEEE Transactions on Geoscience and Remote Sensing* 23: 888-895.

Trenberth, K.E., J.M. Caron, and D.P. Stepaniak. 2001. The atmospheric energy budget and implications for surface fluxes and ocean heat transports. *Climate Dynamics* 17: 259-276.

Tucker, C.J. 1979. Red and photographic infrared linear combinations for monitoring vegetation. *Remote Sensing of Environment* 8(2): 127-150.

Tucker, C.J., J.R.G. Townshend, and T.E. Goff. 1985. African land cover classification using satellite data. *Science* 227: 369-375.

Tucker, C.J., I.Y. Fung, C.D. Keeling, and R.H.Gammon. 1986. Relationship between atmospheric CO_2 variations and a satellite-derived vegetation index. *Nature* 319 (6050): 195-199.

Tucker, C.J., H.E. Dregne, and W.W. Newcomb. 1991. Expansion and contraction of the Sahara desert from 1980 to 1990. *Science* 253: 299-301.

Tung, K.K., M.K.W. Ko, and J.M. Rodriguez. 1986. Are Antarctic ozone variations a manifestation of dynamics or chemistry? *Nature* 333: 811-814.

Turner, B.L., W.C. Clark, R.W. Kates, J.F. Richards, J.T. Mathews, and W.B. Meyer eds. 1990. The Earth as Transformed by Human Action. Cambridge University Press, New York.

Underwood, E., S. Ustin, and D. DiPietro. 2003. Mapping non-native species using hyperspectral imagery. *Remote Sensing of Environment* 86(2): 150-161.

Underwood, E., S.L. Ustin, and C. Ramirez. 2006. A comparison of spatial and spectral image resolution for mapping invasive plants in coastal California. *Ecological Management* 39(1): 63-83.

Ustin, S.L., C.A. Wessman, B. Curtiss, E. Kasischke, J. Way, and V.C. Vanderbilt. 1991. Opportunities for using the EOS imaging spectrometers and synthetic aperture radar in ecological models. *Ecology* 72: 1,934-1,945.

Ustin, S.L., D.A. Roberts, J.A. Gamon, G.P. Asner, and R.O. Green. 2004. Using imaging spectroscopy to study ecosystem processes and properties. *Bioscience* 54: 523-534.

Van Allen, J.A., G.H. Ludwig, E.C. Ray, and C.E. McIlwain. 1958. Observation of high intensity radiation by satellites 1958 Alpha and Gamma. *U.S. National Academy of Sciences, I.G.Y. Satellite Report Series* 3: 73-92.

Van Allen, J.A., and L.A. Frank. 1959. Radiation around the Earth to a radial distance of 107,400 km. *Nature* 183: 430-434.

Van Wilgen, B., M.O. Andreae, J.G. Goldammer, and J.A. Lindesay (eds.). 1997. *Fire in Southern African Savana: Ecological and Atmospheric Perspectives*. University of Witwatersrand Press, Johannesburg.

Von Ahn, J.M., J.M. Sienkiewicz, and P.S. Chang. 2006. Operational impact of QuikSCAT winds at the NOAA Ocean Prediction Center. *Weather Forecasting* 21: 523-539.

Vonder Haar, T.H., and V. Suomi. 1969. Satellite observations of the Earth's radiation budget. *Science* 14: 667-669.

Vonder Haar, T.H., and V. Suomi. 1971. Measurements of the Earth's radiation budget from satellites during a five-year period. Part I: Extended time and space means. *Journal of the Atmospheric Sciences* 28: 305-314.

Vonder Haar, T.H., and A.H. Oort. 1973. New estimate of annual poleward energy transport by northern hemisphere oceans. *Journal of Physical Oceanography* 2: 169-172.

Wahr, J., S. Swenson, V. Zlotnicki, and I. Velicogna. 2004. Time-variable gravity from GRACE: First results. *Geophysical Research Letters* 31: L11501, doi:10.1029/2004GL019779.

Wahr, J., S. Swenson, and I. Velicogna. 2006. The accuracy of GRACE mass estimates. *Geophysical Research Letters* 33: L06401 doi:10.1029/2005GL025305

Wallace, J.M., T.P. Mitchell, and C. Deser. 1989. The influence of sea-surface temperature on surface wind in the eastern equatorial Pacific: Seasonal and interannual variability. *Journal of Climate* 2: 1,492-1,499.

Ware, R., M. Exner, D. Feng, M. Gorbunov, K. Hardy, B. Herman, Y. Kuo, T. Meehan, W. Melbourne, C. Rocken, W. Schreiner, S. Sokolovskiy, F. Solheim, X. Zou, R. Anthes, S. Businger, and K. Trenberth. 1996. GPS sounding of the atmosphere from low Earth orbit: Preliminary results. *Bulletin of the American Meteorological Society* 77: 19-40.

Warren, S. G. 1982. Optical properties of snow. *Reviews of Geophysics and Space Physics* 20(1): 67-89.

Waters, J.W., L. Froidevaux, W.G. Read, G.L. Manney, L.S. Elson, D.A. Flower, R.F. Jarnot, and R.S. Harwood. 1993. Stratospheric ClO and ozone from the Microwave Limb Sounder on the Upper Atmosphere Research Satellite. *Nature* 362: 597-602.

Wayne, R.P. 1985. *Chemistry of Atmosphere*. Clarendon, Oxford.

Weng, F., L. Zhao, G. Poe, R. Ferraro, X. Li, and N. Grody. 2003. AMSU cloud and precipitation algorithms. *Radio Science* 38: 8,068-8,079.

Wexler, H. 1954. Observing the weather from a satellite vehicle. *Journal of the British Interplanetary Society* 13: 269-276.

Wielicki, B.A., T. Wong, R.P. Allan, A. Slingo, J.T. Kiehl, B.J. Soden, C.T. Gordon, A.J. Miller, S.K. Yang, D.A. Randall, F. Robertson, J. Susskind, and H. Jacobowitz. 2002. Evidence for large decadal variability in the tropical mean radiative energy budget. *Science* 295: 841-844.

Williams, A.P., and E.R. Hunt. 2002. Estimation of leafy spurge cover from hyperspectral imagery using mixture tuned matched filtering. *Remote Sensing of Environment* 82: 446-456.

Willson, R.C., S. Gulkis, M. Jansen, N.S. Hudson, and G.A. Chapman. 1981. Observations of solar irradiance variability. *Science* 211: 700-702.

Willson, R.C., and H.S. Hudson. 1988. Solar luminosity variations in solar cycle-21. *Nature* 332: 810-812.

Wilson, M.F., and A. Henderson-Sellers. 1985. Land cover and soils data sets for use in general circulation climate models. *Journal of Climatology* 5: 119-143.

Winker, D.M., R.H. Couch, and M.P. McCormick. 1996. An overview of LITE: NASA's Lidar In-space Technology Experiment. *Proceedings of the Institute of Electrical and Electronics Engineers* 84: 164-180.

Winker, D.M. 1997. LITE: Results, performance characteristics, and data archive. *Proceedings of SPIE* 3218: 186-193.

WMO (World Meteorological Organization). 2006. Antarctic Ozone Bulletin, No. 7. Available at http://www.wmo.int/pages/prog/arep/documents/ant-bulletin-7-2006_000.pdf. Accessed July 12, 2007.

WMO. 2007. Scientific Assessment of Ozone Depletion: 2006. Global Ozone Research and Monitoring Project—Report No. 50. WMO, Geneva, Switzerland. Available at http://ozone.unep.org/Assessment_Panels/SAP/Scientific_Assessment_2006.

WMO-UNEP. 2006. Executive Summary of WMO/UNEP Scientific Assessment of Ozone Depletion: 2006. Available at http://www.wmo.ch/pages/prog/arep/gaw/ozone_2006/ozone_asst_report.html. Accessed July 12, 2007.

Wofsy, S.C., M.B. McElroy, and Y.L. Yung. 1975. The chemistry of atmospheric bromine. *Geophysical Research Letters* 2: 215-218.

Wong, T., B.A. Wielicki, R.B. Lee, G.L. Smith, K.A. Bush, and J.K. Willis. 2006. Reexamination of the observed decadal variability of the earth radiation budget using altitude-corrected ERBE/ERBS nonscanner WFOV data. *Journal of Climate* 19: 4,028-4,040.

Worden, H.M., J. Logan, J.R. Worden, R. Beer, K. Bowman, S.A. Clough, A. Eldering, B. Fisher, M.R. Gunson, R.L. Herman, S.S. Kulawik, M.C. Lampel, M. Luo, I.A. Megretskaia, G.B. Osterman, and M.W. Shephard. 2007. Comparisons of Tropospheric Emission Spectrometer (TES) ozone profiles to ozonesondes: Methods and initial results. *Journal of Geophysical Research* 112: D03309, doi:10.1029/2006GL027806.

Wuebbles, D.J., F.M. Luther, and J.E. Penner. 1983. Effect of coupled anthropogenic perturbations on stratospheric ozone. *Journal of Geophysical Research* 88: 1,444-1,456.

Wunsch, C., and R. Ferrari. 2004. Vertical mixing, energy and the general circulation of the oceans. *Annual Review of Fluid Mechanics* 36: 281-314.

Wunsch, C. 2007. The past and future ocean circulation from a contemporary perspective. *AGU Monograph*, in press.

Wyrtki, K. 1975. El Niño—The dynamic response of the equatorial Pacific Ocean to atmospheric forcing. *Journal of Physical Oceanography* 5: 572-584.

Xie, S.-P. 2004. Satellite observations of cool ocean-atmosphere interaction. *Bulletin of the American Meteorological Society* 85: 195-208.

Yang E.-S., D.M. Cunnold, R.J. Salawitch, M.P. McCormick, J. Russell III, J.M. Zawodny, S. Oltmans, and M.J. Newchurch. 2006. Attribution of recovery in lower-stratospheric ozone. *Journal of Geophysical Research* 111: D17309, doi:10.1029/2005JD006371.

Yung, Y.L., J.P. Pinto, R.T. Watson, and S.P. Sander. 1980. Atmospheric bromine and ozone perturbations in the lower stratosphere. *Journal of Atmospheric Science* 37: 339-353.

Yurganov, L.N., P. Duchatelet, A.V. Dzhola, D.P. Edwards, F. Hase, I. Kramer, E. Mahieu, J. Mellqvist, J. Notholt, P.C. Novelli, A. Rockmann, H.E. Scheel, M. Schneider, A. Schulz, A. Strandberg, R. Sussmann, H. Tanimoto, V. Velazco, J.R. Drummond, and J.C. Gille. 2004. Increased northern hemisphere carbon monoxide burden in the troposphere in 2002 and 2003 detected from the ground and from space. *Atmospheric Chemistry & Physics Discussions* 4: 4,999-5,017.

Zapotocny, T.H., J.A. Jung, J.F. Le Marshall, and R.E. Treadon. 2007. A two-season impact study of satellite and in situ data in the NCEP Global Data Assimilation System. *Weather Forecasting* 22: 887-909.

Zarco-Tejada, P.J., J.R. Miller, G.H. Mohammed, and T.L. Noland. 2000a. Chlorophyll fluorescence effects on vegetation apparent reflectance: I. Leaf-level measurements and model simulation. *Remote Sensing of Environment* 74: 582-595.

Zarco-Tejada, P.J., J.R. Miller, G.H. Mohammed, T.L. Noland, and P.H. Sampson. 2000b. Chlorophyll fluorescence effects on vegetation apparent reflectance: II. Laboratory and airborne canopy level measurements with hyperspectral data. *Remote Sensing of Environment* 74: 596-608.

Zarco-Tejada, P.J., C.A. Rueda, and S.L.Ustin. 2003. Water content estimation in vegetation with MODIS reflectance data and model inversion methods. *Remote Sensing of Environment* 85: 109-124.

Zebiak, S.E. 1982. A simple atmospheric model of relevance to El Niño. *Journal of Atmospheric Science* 391: 2,017-2,027.

Zebker, H. 2000. Studying the Earth with interferometric radar. *Computing in Science and Engineering* 2(3): 52-60.

Zebker, H.A., F. Amelung, and S. Jonsson. 2000. Remote Sensing of Volcano Surface and Internal Processes Using Radar Interferometry. Pp. 179-205 in Mouginis-Mark et al., eds., *Remote Sensing of Active Volcanism*. AGU Monograph No. 116, American Geophysical Union, Washington, DC.

Zhao, H., and R. Shibasaki, 2001, Reconstructing textured CAD model of urban environment using vehicle-borne laser range scanner and line camera. http://lvis.gsfc.nasa.gov/ws_images /abstracts/old/Abstracts_070201.pdf, p.20.

Ziemke, J.R., S. Chandra, and P.K. Bhartia. 1998. Two new methods for deriving tropospheric column ozone from TOMS measurements: Assimilated UARS MLS/HALOE and convective-cloud differential techniques. *Journal of Geophysical Research* 103(D17): 22,115-22,128.

Ziemke, J.R., S. Chandra, and P.K. Bhartia. 2001. "Cloud slicing": A new technique to derive upper tropospheric ozone from satellite measurements. *Journal of Geophysical Research* 106(D9): 9,853-9,867.

Ziemke, J.R., and S. Chandra. 2005. A 25-year data record of atmospheric ozone in the Pacific from Total Ozone Mapping Spectrometer (TOMS) cloud slicing: Implications for ozone trends in the stratosphere and troposphere. *Journal of Geophysical Research* 110: D15105, doi:10.1029/2004JD005687.

Zwally, H.J., J.C. Comiso, C.L. Parkinson, D.J. Cavalieri, and P. Gloersen. 2002. Variability of Antarctic sea ice 1979-1998. *Journal of Geophysical Resarch* 107(C5): 3041, doi:10.1029/2000JC000733.

Appendixes

Appendix A

Examples of Scientific Accomplishments and Relevant Satellite Missions

TABLE A.1 Examples of Landmark Satellite Missions That Have Contributed Significantly to the Scientific Accomplishments Discussed in this Report

Satellite	Accomplishment
ATS/SMS/GOES Meteosat GMS	• Weather observations of the tropics and midlatitudes from geostationary altitude (2, 3) • Tropical cyclone detection and forecasting (3)
Aura	• Observing stratospheric ozone (5) • Observing trace gases in the stratosphere, ozone chemistry (5) • Observing trace gases in the troposphere, tropospheric chemistry and transport (5) • Tropospheric ozone (5) • Global climatology of aerosols (4)
CloudSat CALIPSO	• Global distribution of cloud properties (4)
Envisat	• Stratospheric ozone (5) • Stratospheric trace gases (5) • Tropospheric ozone (5) • Tropospheric trace gases (5)
ERS 1 and 2	• Observing stratospheric ozone (5) • Observing trace gases in the stratosphere (5) • Tropospheric ozone (5) • Observing trace gases in the troposphere (5) • Glacier extent and position of equilibrium line (7) • Understanding ocean tides (8) • Westward-propagating sea surface height variability (8) • Mapping global fires (10) • First images of earthquakes (11)
Explorer 7	• Earth radiation budget (2, 4)
GPS	• Plate tectonics (11)
GRACE	• Analysis of groundwater (6) • Geodesy (11) • Mean Gravity Model (11)
ICESat	• Ice shelf collapse (7)

continued

TABLE A.1 Continued

Satellite	Accomplishment
LAGEOS	• Geodesy (11)
Landsat	• Seasonal snow cover (6) • Increasing growing season (9) • Studying plant physiology (9) • Monitoring agricultural lands (10) • Estimating tropical deforestation (10) • Mapping global land cover (2, 10) • Understanding desertification (10) • Monitoring urban areas (10)
Nimbus series	• Observing stratospheric dynamics (5) • Observing distribution and decrease of stratospheric ozone (5) • Measuring stratospheric trace gases (5) • Antarctic ozone hole (5) • Tropospheric ozone (5) • Declining Arctic summer sea ice (7) • Satellite images corroborate Sverdrup's theory (8) • First global maps of marine primary productivity (9) • First atmospheric soundings (3, 4, 5)
NOAA	• Impact of a volcanic eruption on climate (4) • Global sea surface temperature observations (8) • Atmospheric temperature and moisture soundings for weather prediction (3) • Monitoring global total ozone column over Antarctica (5)
QuikScat	• Ocean wind measurements reveal two-way ocean-atmosphere interaction (8)
RadarSat	• Nonuniform and dynamic ice streams in Antarctica (7) • Declining Arctic summer sea ice (7)
SRTM	• Use of satellite-derived elevation data in hydrology (6) • First fine-resolution topography map (11)
Terra/Aqua	• Distribution of tropospheric carbon monoxide, ozone precursor (5) • Mapping global fires (10) • Indirect effects of aerosols (4) • Global distribution of cloud properties (4) • Global climatology of aerosols (4) • Global marine and terrestrial primary productivity (9) • The carbon cycle (9)
TIROS	• Weather imagery (2, 3) • Numerical weather prediction (3)
TOPEX/Poseidon	• Global mean sea level (8) • Understanding ocean tides (8) • Westward propagating sea-surface height variability (8) • Discovery of internal tides and their contribution to ocean mixing (8) • El Niño (9)
TRMM	• Precipitation over the oceans and the tropics (6) • Ligntning as source of tropospheric NO_x (5)
UARS	• Stratospheric ozone distribution (5) • Role of chloride in ozone depletion (5) • Transport and partitioning of chlorine species in stratosphere (5) • Depiction of tropical tape recorder (5)

NOTE: ERS = European Remote Sensing Satellite; GOME = Global Ozone Monitoring Experiment; GPS = Global Positioning System; GRACE = Gravity Recovery and Climate Experiment; ICESat = Ice, Cloud, and Land Elevation Satellite; NOAA = National Oceanic and Atmospheric Administration; SRTM = Shuttle Radar Topography Mission; TIROS = Television InfraRed Observation Satellite; TOPEX = Topography Experiment; TRMM = Tropical Rainfall Measuring Mission.

Appendix B

Acronyms

ADEOS Advanced Earth Observing Satellite
AgriSTARS Agriculture and Resources Inventory Experiment Through Aerospace Remote Surveys
AIRS Atmospheric Infrared Sounder
ALBH Albert Head
ALOS Advanced Land Observation Satellite (Japan)
AMI Advanced Monitoring Initiative
AMSR-E Advanced Microwave Scanning Radiometer-EOS
AMSU Advanced Microwave Sounding Unit
APAR Absorbed photosynthetic radiation
ARGOS Advanced Research and Global Observation Satellite
ASTER Advanced Spaceborne Thermal Emission and Reflection Radiometer
ASTR Along Track Scanning Radiometer
ATS Applications Technology Satellite
ATSR-M Along Track Scanning Radiometer-Microwave
AVHRR Advanced Very High Resolution Radiometer
AVIRIS Airborne Visible/Infrared Imaging Spectrometer
AWiFS Advanced Wide Field Sensor
B Biomass
BGC BioGeochemical cycles
BOMEX Barbados Oceanographic and Meteorological Experiment
BrO Bromine oxide
BUV Backscattered ultraviolet
CALIPSO Cloud Aerosol Lidar and Infrared Pathfinder Satellite Observations
CASA Carnegie-Ames-Stanford Approach
CFC Chlorofluorocarbon
CHAMP Challenging Mission Payload
ClO Chlorine monoxide
$ClONO_2$ Chlorine nitrate
CloudSat Cloud Satellite
CNES Centre Nationales d'Etudes Spatiales
CO Carbon monoxide
CO_2 Carbon dioxide
COSMIC Constellation Observing System for Meteorology, Ionosphere, and Climate
CRS Cloud Radar System
CSA Canadian Space Agency
CZCS Coastal Zone Color Scanner
DLR Deutsches Zentrum für Luft—und Raumfahrt
DMSP Defense Meteorological Satellite Program
DORIS Doppler Orbitography and Radiopositioning Integrated by Satellite
DU Dobson units
ECMWF European Centre for Medium Range Forecasting
ENSO El Niño-Southern Oscillation
ENVISAT European Enivronmental Satellite
EORC Earth Observation Research Center
EOS Earth Observing System
ERBE Earth Radiation Budget Experiment
ERS European Remote Sensing Satellite
ERTS Earth Resources Technology Satellite
ESA European Space Agency
ESMR Electrically Scanning Microwave Radiometer
ETM Enhanced Thematic Mapper
FAO Food and Agriculture Organization
FEWS NET Famine Early Warning System Network
FGGE First GARP Global Experiment
FORMOSAT Formosa Satellite
GAC Global Area Coverage
GARP Global Atmospheric Research Program
GATE GARP Atlantic Tropical Experiment
GBA Global Burnt Area

GCM	Global Climate Model
GDR	Geophysical Data Record
GEO	Geosynchronous orbit
Geos3	Geodynamics Experimental Ocean Satellite 3
GIS	Geographic information system
GLC	Global Land Cover
GLCC	Global Land Cover Classification
GLIMS	Global Land Ice Measurement from Space
GLOBSCAR	Global Burn Scar
GMCC	Global Monitoring for Climate Change
GMS	Geostationary Meteorological Satellite
GOES	Geostationary Operational Environmental Satellites (Japan)
GOME	Global Ozone Monitoring Experiment
GOMS	Geostationary Operational Meteorological Satellite (Russia)
GPM	Global Precipitation Mission
GPS	Global Positioning System
GPS/MET	Global Positioning System/Meteorology
GRACE	Gravity Recovery and Climate Experiment
GSFC	Goddard Space Flight Center
GWE	Global Weather Experiment
HALOE	Halogen Occultation Experiment
HCl	Hydrochloric acid
HIRS	High Resolution Infrared Radiation Sounder
HNLC	High nutrient low chlorophyll
I	Light intensity
ICESat	Ice, Cloud, and Land Elevation Satellite
IGY	International Geophysical Year
InSAR	Interferometric synthetic aperture radar
IPAR	Intercepted photosynthetically active radiation
IR	Infrared
IRIS	Infrared Interferometer Spectrometer
IRS	Indian Remote Sensing Satellite
ISCCP	International Satellite Cloud Climatology Project
ITOS	Improved TIROS Operational System
JAXA	Japanese Aerospace Exploration Agency
JERS	Japanese Earth Resources Satellite
JPL	Jet Propulsion Laboratory
LAC	Local area coverage
LACIE	Large Area Crop Inventory Experiment
LAGEOS	Laser Geodynamics Satellites
LAI	Leaf area index
Landsat	Land Satellite
LaRC	Langley Research Center
LEO	Low earth orbit
Lidar	Light detection and ranging
LIMS	Limb Infrared Monitor of the Stratosphere
LRIR	Limb Radiance Inversion Radiometer
MABL	Marine atmospheric boundary layer
MAPS	Measurement of Air Pollution from Satellites
MCC	Mesoscale Convective Complex
McIDAS	Man-Computer Interactive Data Access System
MERIS	Medium Resolution Imaging Spectrometer
MET	Meteorology
MISR	Multiangle Imaging Spectroradiometer
MLS	Microwave Limb Sounder
MODIS	Moderate Resolution Imaging Spectroradiometer
MOPITT	Measurement of Pollution in the Troposphere
MOS	Maritime Observation Satellite
MSS	Multispectral scanner
MTSAT	Multi-functional Transport Satellite (Japan)
NAE	National Academy of Engineering
NAS	National Academy of Sciences
NASA	National Aeronautics and Space Administration
NCAR	National Center for Atmospheric Research
NCEP	National Centers for Environmental Prediction
NDVI	Normalized difference vegetation index
NHC	National Hurricane Center
NIR	Near-infrared
NOAA	National Oceanic and Atmospheric Administration
NO_x	Nitrogen oxides
NPP	Net primary productivity
NRC	National Research Council
NSCAT	NASA Scatterometer
NSIDC	National Snow and Ice Data Center
NWS	National Weather Service
O_3	Ozone
OCI	Ocean Color Imager
OCTS	Ocean Color and Temperature Sensor
OMI	Ozone Monitoring Instrument
P	Productivity
PECAD	Production Estimates and Crop Assessment Division
P-I	Photosynthesis irradiance
PR	Precipitation radar
PRI	Photochemical reflective index
PSC	Polar stratospheric clouds
QuikSCAT	Quick Scatterometer
RA	Radar Altimeter
Radar	Radio detection and ranging
RAPID	Research into Adaptive Particle Imaging Detectors
RBV	Return beam videocon
RES	Radio-echo sounding
SAGE	Stratospheric Aerosol and Gas Experiment
SAMS	Stratospheric and Mesospheric Sounder
SAR	Synthetic aperture radar
SBUV	Solar backscattered ultraviolet

SCIAMACHY	Scanning Imaging Absorption Spectrometer for Atmospheric Chartography
SCLP	Snow and Cold Land Processes
SCR	Selective Chopper Radiometer
SeaWiFS	Sea-Viewing Wide Field-of-View Sensor
SIR	Shuttle Imaging Radar
SIRS	Satellite Infrared Spectrometer
SMMR	Scanning Multichannel Microwave Radiometer
SMS	Synchronous Meteorological Satellite
SO_2	Sulfur dioxide
SPOT	Satellite Probatoire pour l'Observation de la Terre
SRTM	Shuttle Radar Topography Mission
SSEC	Space Science and Engineering Center
SSH	Sea surface height
SSM/I	Special Sensor Microwave/Imager
SSM/R	Special Sensor Microwave/Radiometer
SST	Sea surface temperature
TIROS	Television Infrared Observation Satellite
TM	Thematic Mapper
TM/ETM	Thematic Mapper/Enhanced Thematic Mapper
TMI	TRMM Microwave Imager
TOMS	Total Ozone Mapping Spectrometer
TOPEX	Topography Experiment
T/P	TOPEX/Poseidon
TRMM	Tropical Rainfall Measuring Mission
TSI	Total solar irradiance
TW	Threat Warning
TWS	Terrestrial Water Storage
UARS	Upper Atmosphere Research Satellite
UNCCD	United Nations Convention to Combat Desertification
USDA	U.S. Department of Agriculture
UTC	Coordinated Universal Time
UV	Ultraviolet
VCF	Vegetation Continuous Fields
VGPM	Vertically Generalized Production Model
VIRS	Visible Infrared Scanner
VISSR	Visible and Infrared Spin Scan Radiometer
WINDSAT	Wind Satellite
W/m^2	Watts per square meter

Appendix C

Biographical Sketches of Committee Members and Staff

COMMITTEE MEMBERS

J. Bernard Minster (*Chair*) is a professor of geophysics at the Institute of Geophysics and Planetary Physics of the Scripps Institution of Oceanography, University of California, San Diego (UCSD), and senior fellow at the San Diego Supercomputer Center. For the past year he has been chair of the UCSD Division of the University of California Academic Senate. Dr. Minster's research interests are centered on the determination of the structure of the Earth's interior from broadband seismic data, by imaging the Earth's upper mantle and crust using seismic waves. This research has led him to an involvement in the use of seismic means for verification of nuclear test ban treaties. He has long been interested in global tectonic problems and in the application of space-geodetic techniques, including synthetic aperture radar and laser altimetry, to study tectonic and volcanic deformations of the Earth's crust by airborne and spaceborne remote sensing. He is a member of the ICESat science team, which uses the GLAS instrument to measure ice-sheet mass balance and global topographic change. He has been principal investigator on several proposed SAR missions in low Earth orbit and on a proposed laser altimetry mission to Europa. More recently he has led major efforts toward estimating the effects of very large earthquakes in southern California, using supercomputer simulations, and analyzing paleoseismic data using hyperspectral imaging. He has held positions in industry and has been a consultant and reviewer for numerous companies. He was the Nordberg Lecturer at NASA GSFC in 1996 and was elected a fellow of the American Geophysical Union in 1990. He is chair of the recently created Earth and Space Science Informatics Focus Group of the American Geophysical Union. Dr. Minster has chaired previous National Research Council committees, including the Committee on Geophysical and Environmental Data, and has served on numerous committees related to solid earth geophysics, including the Board on Earth Sciences and Resources and its Committee on Geodesy. He is currently vice-chair of the World Data Center Panel of the International Council of Scientific Unions.

Janet W. Campbell (*Vice Chair*) is director of the Center of Excellence for Coastal Ocean Observation and Analysis, which was established in August 2002 as part of NOAA's Coastal Observation Technology System. Dr. Campbell is also director of the Ocean Process Analysis Laboratory, one of four centers that comprise the Institute for the Study of Earth, Oceans, and Space at the University of New Hampshire (UNH). Dr. Campbell is a member of NASA's Ocean Color Science Team and the MODIS Instrument Team. She has been at the University of New Hampshire since 1993 and is a member of the graduate faculty in the Earth Sciences Department. Between 1997 and 1999 she served as program manager for ocean biology and biogeochemistry at NASA headquarters in Washington, D.C. Before joining UNH, she was a research scientist at the Bigelow Laboratory for Ocean Sciences in Boothbay Harbor, Maine (1982-1993), where she established and directed the remote sensing computer facility. She previously worked as an aerospace technologist/engineer at the NASA Langley Research Center in Hampton, Virginia. She holds a Ph.D. in statistics from Virginia Polytechnic Institute and a master's degree in mathematics from Vanderbilt University. She previously served on the National Research Council Committee on Earth Studies and two other NRC studies.

Jeff Dozier is a professor in the Donald Bren School of Environmental Science and Management at the University of California, Santa Barbara. He founded the Bren School and served as its first dean for 6 years. Dr. Dozier earned his Ph.D. in geography from the University of Michigan. His research interests are in the fields of snow hydrology, Earth system science, remote sensing, and information systems. He has pioneered interdisciplinary studies in two areas:

one involves the hydrology, hydrochemistry, and remote sensing of mountainous drainage basins; the other is in the integration of environmental science and remote sensing with computer science and technology. He was the senior project scientist for NASA's Earth Observing System when the configuration for the system was established. He also helped found the MEDEA group, which investigated the use of classified data for environmental research, monitoring, and assessment. Dr. Dozier has chaired or served on numerous NRC committees concerned with data and computational sciences, and he is the current chair of the Committee on Geophysical and Environmental Data. He is a fellow of the American Geophysical Union, the American Association for the Advancement of Science, and the UK's National Institute for Environmental eScience. He is also an honorary professor of the Academia Sinica and a recipient of the NASA/Department of Interior William T. Pecora Award and the NASA Public Service Medal.

James R. Fleming is professor of science, technology and society at Colby College. His research interests include the history of the geophysical sciences, especially meteorology, climatology, and oceanography. Professor Fleming earned a B.S. in astronomy from Pennsylvania State University, an M.S. in atmospheric science from Colorado State University, and an M.A. and a Ph.D. in the history of science from Princeton University. He is founder of the International Commission on History of Meteorology and editor of its journal, *History of Meteorology*. In 2003 Professor Fleming was elected a fellow of the American Association for the Advancement of Science (AAAS) "for pioneering studies on the history of meteorology and climate change and for the advancement of historical work within meteorological societies." He was also awarded the Ritter Memorial Fellowship at Scripps Institution of Oceanography. In 2005-2006 Professor Fleming held the Charles A. Lindberg Chair in Aerospace History at the Smithsonian's National Air and Space Museum and in 2006-2007 he was awarded the Roger Revelle Fellowship in Global Stewardship from the AAAS.

John C. Gille is a researcher at the National Center for Atmospheric Research and the University of Colorado, studying chemical processes and their impact on climate and air quality. He applies his training as a physicist to the development and use of remote sensing instruments to study the chemical composition, dynamics, and transport of trace species in the troposphere and middle atmosphere. At present he serves as the U.S. principal investigator for MOPITT, an instrument flying aboard NASA's Terra spacecraft that measures the global distributions of carbon monoxide in the troposphere. Dr. Gille is also the U.S. principal investigator for the High Resolution Dynamics Limb Sounder (HIRDLS), an instrument on NASA's Aura satellite that scientists use to study the ozone layer, climate change, and more. He was previously the principal investigator for the LRIR that flew on Nimbus 6 and LIMS on Nimbus 7 and was a collaborative investigator on the Cryogenic Limb Array Etalon Spectrometer (CLAES) on the Upper Atmosphere Research Satellite. He is a fellow of the American Geophysical Union and the American Meteorological Society. Dr. Gille has served on two previous committees for the NRC, the Committee on Data Management and Computation and the Committee on Earth Studies.

Dennis L. Hartmann joined the faculty of the University of Washington in 1977 and is currently professor and chair of the Department of Atmospheric Sciences, adjunct professor of the Quaternary Research Center, senior fellow of the Joint Institute for the Study of the Atmosphere and Ocean, and a former member of the Board of Directors of the Program in Climate Change. Dr. Hartmann received his Ph.D. in geophysical fluid dynamics from Princeton University in 1975. Dr. Hartmann has published more than a hundred articles in referred scientific journals and published a textbook, *Global Physical Climatology*, in 1994. His primary areas of expertise are atmospheric dynamics, radiation and remote sensing, and mathematical and statistical techniques for data analysis. He is a fellow of the American Meteorological Society, the American Geophysical Union, and the American Association for the Advancement of Science. Dr. Hartmann recently chaired a study of climate feedback processes for the National Research Council and is currently a Board on Atmospheric Sciences and Climate board member. He has also served on numerous advisory, editorial, and review boards for the NRC, National Science Foundation, NASA, and NOAA.

Kenneth Jezek is a professor in the Byrd Polar Research Center and the School of Earth Sciences at The Ohio State University. He is the principal investigator of the RADARSAT Antarctic Mapping Project. His research interests include remote sensing studies of sea ice and the polar ice caps, including applications of synthetic aperture radar data to gauge the response of ice sheets to changing climate. Dr. Jezek is a past member of the NRC Committee for Review of the Science Implementation Plan of the NASA Office of Earth Science, the Scientific Committee on Antarctic Research, the Committee on Glaciology, and the Panel on Climate Variability and Change.

Stan Kidder is a senior research scientist at the Cooperative Institute for Research in the Atmosphere at Colorado State University. Dr. Kidder received his Ph.D. in atmospheric science from Colorado State University in 1979. His research centers on application of satellite data to meteorological problems. He is also studying the blending of products produced from different sensors on different satellites into unified products and the development of new orbits and constellations for future meteorological satellites. Dr. Kidder was the co-lead instructor for the COMET SatMet course and is the author (with T. H. Vonder Haar) of *Satellite Meteorology: An Introduction* (Academic Press, 1995). He has been a member of numerous committees, including the American Meteorological Society Board on Higher Education.

Navin Ramankutty signed on as assistant professor in the Department of Geography at McGill University in June 2006. Previously, he was an assistant scientist at the Center for Sustainability and the Global Environment (SAGE) at the University of Wisconsin. Dr. Ramankutty joined SAGE as a research scientist in May 2000 and led its efforts on documenting contemporary and historical patterns of land use and land cover across the world. Working with colleagues at SAGE, Dr. Ramankutty developed a statistical data fusion technique to merge satellite data and socioeconomic data on agricultural land use to develop a global data set of the world's croplands. Dr. Ramankutty and his team have further developed global agricultural land-use data sets, focusing on more detailed characterizations of the world's major crops, their yield and production, and farming practices. These emerging data sets have become extremely popular with the global change community. They have attracted widespread media attention, including becoming part of a National Geographic Society pullout map in September 2002 and being used in the 8th edition of the *National Geographic Atlas of the World*.

Anne M. Thompson is a professor in the Department of Meteorology at the Pennsylvania State University with research interests in atmospheric chemistry: modeling and measurements of trace gases, air-sea gas exchange, biomass burning, and remote sensing. As co-mission scientist for NASA's 1997 DC-8 SONEX (SASS Ozone and Nitrogen Oxides Experiment), Dr. Thompson demonstrated that lightning, convection, and aircraft emissions have comparable perturbations in the North Atlantic upper troposphere. Since 1998, Dr. Thompson has been principal investigator for SHADOZ (Southern Hemisphere Additional Ozonesondes), analyzing tropical ozonesonde data for satellite validation and climate studies. She also led the 2004 and 2006 INTEX Ozonesonde Network Study campaigns in the first strategic ozonesonde sampling over North America. Dr. Thompson is a fellow of the American Meteorological Society, the American Association for the Advancement of Science, and the American Geophysical Union. She has been awarded the Committee on Space Research (COSPAR) Nordberg Medal for space science and the Women in Aerospace International Achievement Award.

Susan L. Ustin is a professor in the Department of Land, Air, and Water Resources at the University of California, Davis (UC Davis). Dr. Ustin received her Ph.D. in botany from UC Davis, in the area of plant physiological ecology. Her multidisciplinary environmental research focuses on developing applications of remote sensing data to assess environmental processes. She began working with the Jet Propulsion Laboratory during the initial stages of NASA's imaging spectrometry program and has since worked extensively with hyperspectral imagery for quantitative plant and soil measurements. She has been a principal investigator and science team member of several NASA sensor programs for Earth observation and is currently a member of the MODIS science team. Dr. Ustin recently served as director of the California Space Institute Center of Excellence at UC Davis and as director of the Western Regional Center for Global Environmental Change. She is the editor of the *Manual of Remote Sensing, Vol. 4, Remote Sensing for Natural Resource Management and Environmental Monitoring* (John Wiley & Sons, 2004). She has served previously on four NRC committees.

James Yoder joined the staff at the Skidaway Institute of Oceanography in Georgia in 1978 and from 1986 to 1988 was a visiting senior scientist at the Jet Propulsion Laboratory assigned to NASA headquarters. He joined the faculty at the Graduate School of Oceanography (GSO) at the University of Rhode Island in 1989 and was promoted to professor in 1992. He was named associate dean of oceanography at GSO in 1993 and served in that capacity until 1998. From 2000 to 2001, Dr. Yoder served as GSO interim dean before moving to the National Science Foundation, where he served as director of the Division of Ocean Sciences from 2001 to 2004 before returning to GSO in October 2004. In November 2005, Dr. Yoder was named vice president for academic programs and dean at the Woods Hole Oceanographic Institution. His research involves the study of oceanographic processes primarily using satellite radiometers observing the ocean in the visible/near-IR wavelengths, including NASA's CZCS, Japan's OCTS, and NASA's SeaWiFS and MODIS. Dr. Yoder lists 90 scientific and other publications and holds a Ph.D. degree in oceanography from the University of Rhode Island. He served on the NRC's Committee on Oceanic Carbon (1992-1994) and was a member of the Ocean Studies Board (2001).

NRC STAFF

Claudia Mengelt is a program officer for the Board on Atmospheric Sciences and Climate (BASC). After completing her B.S. in aquatic biology at the University of California, Santa Barbara, she received her M.S. in biological oceanography from the College of Oceanic and Atmospheric Sciences at Oregon State. Her master's degree research focused on how chemical and physical parameters in the surface ocean effect Antarctic phytoplankton species composition and consequently impact biogeochemical cycles. She obtained her Ph.D. in marine sciences from the University of California, Santa Barbara, where she conducted research on the photophysiology of harmful algal species. She joined the full-time staff of BASC in the fall of 2005 following a fellowship with the NRC Polar Research Board in the winter of 2005. At the National Academies, she has worked on studies addressing the design of Arctic observing systems, providing strategic guidance to NSF's atmospheric sciences, and evaluating lessons learned from global change assessments.

Maria Uhle has been a program officer with the Polar Research Board at the National Research Council since April of 2005. Prior to joining the NRC, she was the Jones Assistant Professor of Environmental Organic Geochemistry in the Department of Earth and Planetary Sciences at the University of Tennessee (UT). At UT, Dr. Uhle mentored several graduate students in various scientific disciplines including Quaternary climate studies, salt marsh ecology, reconstruction of biomass burning events throughout geologic history, organic contaminate remediation and Antarctic biogeochemistry. Dr. Uhle received her B.S. from Bates College, M.S. from the University of Massachusetts, and Ph.D. from the University of Virginia. At the NRC, she has directed several studies including Assessment of the U.S. Coast Guard Polar Icebreakers Roles and Future Needs, Exploration of Antarctic Subglacial Aquatic Environments: Environmental and Scientific Stewardship. She continues to work with the U.S. National Committee on the International Polar Year developing interagency communications and public outreach and education projects.

Leah Probst is a research associate with the NRC's Board on Atmospheric Sciences and Climate and Polar Research Board. Since joining the NRC staff in 1999, Ms. Probst has led studies on the science and implementation plan for the World Climate Research Programme's Americas Prediction Project, on the proposed Global Precipitation Measurement satellite mission, and on stratospheric ozone recovery and its implications for ultraviolet radiation exposure. She works with the U.S. National Committee on the International Polar Year 2007-2008 and with the NRC's Climate Research Committee. She has contributed to many other NRC studies, including topics such as surface temperature reconstructions for the last 2,000 years, the Tropical Rainfall Measuring Mission satellite program, the New Source Review Program of the Clean Air Act, and cumulative effects of oil and gas activities on Alaska's North Slope. She received a B.A. in biology from George Washington University.

Katherine Weller is a senior program assistant for the Board on Atmospheric Sciences and Climate and the Polar Research Board. Since joining the National Academies in 2006, Ms. Weller has worked with the Climate Research Committee, and has worked with committees to review the Climate Change Science Program's Synthesis and Assessment Products 2.4, 3.3, and 5.2. In 2004 she received a B.S. in biopsychology from the University of Michigan. She is currently working toward a master's degree in environmental science and policy from Johns Hopkins University.